Nanotechnologies

Nanotechnologies

Concepts, Processing and Applications

Pierre Camille Lacaze

Series Editor
Pierre-Noël Favennec

iSTE

WILEY

First published 2013 in Great Britain and the United States by ISTE Ltd and John Wiley & Sons, Inc.

ISTE Ltd
27-37 St George's Road
London SW19 4EU
UK

www.iste.co.uk

John Wiley & Sons, Inc.
111 River Street
Hoboken, NJ 07030
USA

www.wiley.com

Library of Congress Control Number: 2012949499

British Library Cataloguing-in-Publication Data
A CIP record for this book is available from the British Library
ISBN: 978-1-84821-438-5

MIX
Paper from responsible sources
FSC® C013604

Printed and bound in Great Britain by CPI Group (UK) Ltd., Croydon, Surrey CR0 4YY

Table of Contents

Preface

Lately, nanotechnologies have become an important social issue. Every so often they are the subject of major debates in the media, and their illustration is sometimes close to science fiction.

It is true that this domain constitutes a huge break with the daily microscopic world to which we are accustomed, even if only by using concepts of quantum mechanics, which are very different from those of classical physics. This leads us to think about physical phenomena in a completely different way.

Effects owing to the minuteness of objects often result in unexpected discoveries, which are increasing in number every day, to which innovative applications are constantly being added. These should revolutionize our way of life in the future.

Nanosciences and nanotechnologies have gradually been introduced, first in a modest and anonymous fashion in the 1960s (though the word "nanotechnology" was used for the first time in 1970), becoming a strategic research principle by the turn of the century. Nanotechnology is now a major field in science that has been developed by industries and universities, encouraged and subsidized by the most influential research organizations of all the developed countries and also every so often used as a form of advertising.

There is no doubt that nanotechnologies are interesting and we are all aware that future developments must come from mastering them. This is a booming industry, in terms of diversity as much as in the number of applications; one day's progress is very rapidly surpassed by discoveries made the day after. Nanotechnologies are a scientific revolution whose implementations have still not ceased to surprise us. Sometimes unpredictable applications appear as a result, and so may be full of promise for improving our way of life in the future, with advantages that override the risks ultimately created by manipulating nanomaterials.

This book realistically presents the ongoing research and the most recent and significant achievements in the field of nanotechnology. The examples chosen are cited with their bibliographic references in a way that allows anyone to refer to the original sources; an overview of the cited article can always be accessed on the Internet.

This book is aimed at knowledgeable readers, and at the same time, those who are curious about how science is progressing or those who wish to inform themselves more accurately about the perspectives nanotechnologies may offer. Physical phenomena are generally outlined in simplified language and are deprived of all mathematical formalism, with supporting boxes of text designed to not only alleviate discussions but also draw the readers' attention to several key points essential for their understanding. There are eight chapters in this book, which are separated into two parts; one describes the main concepts and the other describes the major areas in which nanotechnologies are applied.

Part 1: Concepts, discoveries and the rapid development of nanotechnologies

Chapter 1 describes the emergence of ideas and the incentives of research organizations toward nanotechnologies.

Chapter 2 shows how the discoveries of new carbonaceous nanomaterials with uncommon properties and the invention of new methods of observing the infinitesimal have made nanosciences and nanotechnologies stand out in industry and research.

Chapter 3 describes the most illustrative nano-objects of various shapes and structures and their implementation into structured networks.

Chapter 4 analyzes properties due to confinement of matter in the case of metallic insulating nanomaterials and semiconductors in various forms (nanoparticles, nanowires, etc.), either grouped into networks or ungrouped.

Part 2: Applications and societal implications of nanotechnology

Chapter 5 recounts the evolution of electronics and new orientations toward molecular electronics and spintronics (a new discipline combining magnetism and electronics).

Chapter 6 shows how using nanomaterials can improve yields of stored energy along with the conversion of solar energy into electricity, resulting in important ecological applications to habitats and totally electric vehicles.

Chapter 7 describes recent uses of nanotechnologies in medicine and biology, and the opportunity to update applications for new non-invasive therapies.

Chapter 8 provides an insight into future ideas developed during the design of nanomachines, with examples of promising applications of molecular valves in medicine.

Pierre Camille Lacaze
October 2012

Acknowledgments

The idea of writing this book arose during a lecture given at the Carré Magique in Lannion, convened by ABRET (Breton Research and Technology Association) and APAST (Scientific Animation Association of Trégor). The enthusiasm and remarkable scientific expertise of all members of these committees deserve high appreciation.

I would like to thank several people who have helped and encouraged me to write this book.

I would like to thank P.N. Favennec, series editor of ISTE Ltd and secretary of the scientific board of ABRET, who encouraged me to formalize in writing the contents of the lecture that I gave on nanotechnologies, and also M. Urien, former engineer at CNET (National Centre of Telecommunication Studies) and vice president of APAST, for his indirect contribution to the content of this book and warm welcome shown by him and his wife toward me.

I would also like to thank my laboratory colleagues at ITODYS (Laboratory of Interfaces, Processing, Organization and System Dynamics) at the University Paris Diderot for enabling me to benefit from using the facilities and for privileged online access to all scientific literature.

I would especially like to thank J.-C. Lacroix, Professor at the University Paris Diderot, for his advice and fruitful exchanges that we have had throughout the writing of this book, as well as J.M. El Hage Chahine, director of research at CNRS, and M. Hémadi, lecturer, for their pertinent advice and suggestions about biological problems.

Finally, I would like to thank my wife for helping me with proofreading and for her patience, as well as my children for their encouragement throughout the writing of this book.

PART 1

Concepts, Discoveries and the Rapid Development of Nanotechnologies

Chapter 1

Nanotechnologies in Context: Social and Scientific Awareness of their Impact

1.1. Feynman, the visionary

The term "nanotechnology" was coined in the 1970s by the Japanese professor Norio Taniguchi [TAN 74] to design "infinitesimal" technologies; however, interest in such technologies had already been around well before that time.

Richard Feynman (a physicist famous for his developments in quantum electrodynamics, Nobel Prize awarded 1965) was the first to anticipate the future interest in studying and using properties of objects or systems composed of only a small number of atoms or molecules (atomic or molecular aggregates) to develop new fields of applications based on the practice of a new physical chemistry [FEY 59].

In a lecture given in 1959 at the American Society of Physics congress at the California Institute of Technology (Caltech), he defined a new type of chemistry in which atoms are individually manipulated with the prospect of developing devices at the nanometer scale (a billionth of a meter).

The title of his lecture "There's Plenty of Room at the Bottom" became famous. It was a real incentive for scientists to explore the world of the infinitesimal that he presented as a new El Dorado for science and technology, and for the benefit of society.

Figure 1.1. *Feynman with his students in the 1950s (courtesy of Caltech)*

To illustrate his point and to stir the imagination of the audience, he showed, after a simple computation, that the 25 volumes of *Encyclopedia Brittanica* could be inscribed onto the head of a pin.

Speaking in the context of the 1950s could be somewhat perplexing, considering the advancing state of technology in the field of miniaturization in general, especially that of electronics.

Indeed, we must remember that the invention of the transistor by the American scientists, Shockley, Bardeen and Brattain in Bell laboratories (Nobel Prize awarded 1957) dates back to 1947 [SHO 56], and that the first integrated circuit (an important simplification of electronic circuits) invented by Kilby (Nobel Prize awarded 2000) in Texas Instruments laboratories in the United States, only appeared a decade later, and was far from miniature [KIL 00].

It was not until the early 1970s that the first processors appeared; devices including a very large number of transistors arranged in thin layers on the same base of silicon. We can rightly consider this to have marked the actual beginning of the miniaturization of electronic circuits.

This miniaturization of circuits arranged in thin layers could not have been carried out if it was not for the considerable progress made in the chemistry and physics of surfaces and without the development of instruments for analyzing and controlling them at the microscopic scale.

Figure 1.2. *Shockley, Bardeen and Brattain (from left to right), Bell Labs, 1947*

Yet this is still far from manipulations at the nanometer scale. As a guide, the first microprocessors built in 1971 by Intel® (4004-86 Processor) already containing a decent number of transistors (2,300), of miniscule sizes (approximately 10 μm), are in no way comparable to the most recent processors made in 2010, involving about a billion transistors, and a hundred times smaller in size [INT 2011].

This nicely shows how in the 1960s technology was deeply rooted in the micrometer scale, and that Feynman's premonition could appear more futuristic to some and somewhat utopic to others.

1.2. Nanotechnologies and their definition

After numerous discussions, a consensus emerged in the scientific community around 2000; the field of nanotechnology was rather generally defined as the manipulation of objects with dimensions between 1 and 100 nm. The term "nanotechnologies" is intentionally recommended rather than "nanosciences" in order to emphasize all the social implications that nanotechnologies might be able to bring in the near future.

This definition, *a priori* restrictive, implies that all manipulations of objects with dimensions of nanometers (groups of atoms or molecules) translate as an insertion or a link to macroscopic systems (objects with dimensions of micrometers or millimeters) that are also a part of nanosciences.

So, any modification made to a material, by grafting atoms or molecules onto its surface, or even by incorporating nanosized particles into the core of the material is still found under this definition. In a more general fashion, this comes back to defining these modifications as a restructuring of matter at the nanometer scale. This definition also applies to manipulating complex macroscopic objects to which nanosized active components are associated.

From a practical point of view, defining nanotechnologies according to units in nanometers is rather allusive if we do not have some indications of known objects.

The nanometer, equal to a billionth of a meter (one millionth of a millimeter or one thousandth of a micron) is not a common unit of measurement. This is why, in order to better appreciate this dimension, we have placed (Figure 1.3) some more familiar objects (atoms, molecules, living cells, wavelengths of visible light, transistors and microprocessors from the previous generation, etc.) on to a logarithmic scale.

So, at the lower limits of the scale, between 0.1 and 1 nm, we find atoms (made up of a nucleus and an electron cloud); next come molecules made up of assemblies of atoms whose size clearly reflects the number of atoms.

In the case where the number of atoms making up a molecule is very large, the diameter of the molecule is still less than 1 nm, but its length can vary from several nanometers to tens of nanometers (macromolecule).

In extreme cases such as deoxyribonucleic acid (DNA), a molecule made up of an assembly of two strands of DNA (two macromolecules of complementary structures that spontaneously combine as a result of weak forces), the diameter is less than 1 nm, but the length of the molecule is considerable and can reach up to several hundred nanometers, or even several tens of centimeters (Stryer [STR 95]).

The electronics industry, where remarkable developments have been made in terms of miniaturization, is also at the heart of nanotechnology.

For instance, the last generation of Intel® field-effect transistors (the FET "Tri Gate" is approximately 100 nm), which intervenes in the operating of microprocessors (cm dimensions), could only be carried out thanks to advances in vacuum-deposition techniques, which now allow extremely complex circuits to be made from diverse materials with layers of only a few nanometers thick.

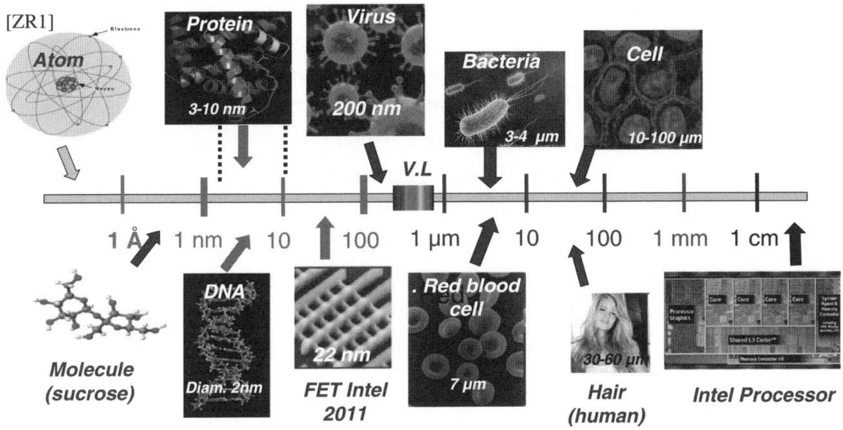

Figure 1.3. *Scale of dimensions of several familiar objects positioned on a logarithmic scale (each interval corresponds to a multiple of 10). The Angström (Å), generally used to define dimensions of atoms, is equal to 0.1 nm. For a color version of this figure see www.iste.co.uk/lacaze/nano.zip*

Beyond 100 nm, the visible light spectrum (VL) is also a familiar dimension, with bands of wavelengths λ extending from 400 nm (violet) to 700 nm (red).

It is also within this range of dimensions that viruses are found (between 200 and 400 nm), whereas bacteria, red blood cells and living cells are found within the micrometer domain, with sizes ranging from several microns to several tens of microns (1 µm = 1,000 nm = 0.001 mm).

Much further in this same range of dimensions, we find the human hair, with thicknesses varying according to the individual between 30 and 80 µm (i.e. 30,000 and 80,000 nm).

Finally, in the centimeter range, microprocessors, comprising an assembly of a considerable number of transistors (around one billion in 2010), which can be regarded as pure products of nanotechnology, presented in the form of a small circuit board of several centimeters.

1.3. The consideration of nanotechnologies by scientific organizations

In the 1960s and the 1970s, a large portion of experiments in physics, chemistry and biology could be considered retrospectively as being within the field of nanoscience, even though this term had not yet been coined.

Given the importance of new applications related to the discovery and study of new materials of nanometric dimensions such as aggregates of atoms or nanoparticles (NPs) or carbon nanomaterials (fullerenes in 1980, nanotubes of carbon in 1985) and also, after new observation instruments in the nanometric scale were invented (near-field scanning microscopy), it was not until the 1990s that public research organizations began to realize the importance of the domain and encouraged the majority of governments to financially support R&D initiatives within nanosciences. It was in the United States that the incentive was greatest. At the initiative of the NSF (National Science Foundation), there was a considerable mobilization among the main components of society, in order to formalize the launch of a vast research program dedicated to all forms of nanotechnologies.

This is the famous National Nanotechnology Initiative (NNI) developed at the beginning of 2000 under the auspices of societies' major scientists and with the support of federal departments.

This NNI establishes a strategic preliminary plan of 10 years for the intensive development of nanotechnologies.

Among the main actions of R&D to be undertaken, there are nine key themes:

1) the nanostructuration of materials;

2) industrial production at the nanoscale;

3) chemical, biological and radiological detection and protection against explosives;

4) instrumentation and metrology;

5) nanoelectronics, nanophotonics and nanomagnetics;

6) health: diagnosis and therapies;

7) conversion of energy and its storage;

8) micromechanics and robotics;

9) upgrading and developing chemical processes at the nanoscale for improving environmental conditions (Roco [ROC 11]).

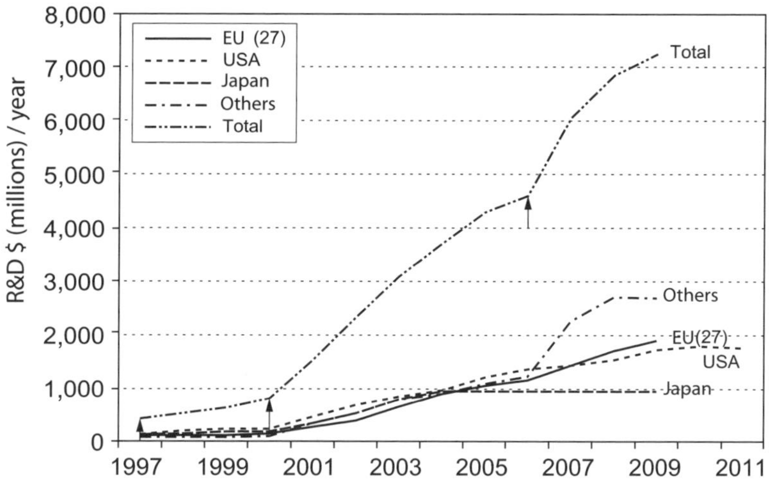

Figure 1.4. *Annual evolution of credits dedicated to R&D in different parts of the world. Notice the acceleration of financing after 2000 and 2006, with a global amount of more than $7 billion in 2009. (China, Korea and Taiwan represent the countries in the category "Others".) (adapted from Roco [ROC 11]). For a color version of this figure see www.iste.co.uk/lacaze/nano.zip*

To highlight how important this plan is, President Clinton personally offered his support in a speech that he gave on January 21, 2000, at Caltech, more than 40 years after that of Feynman [CLI 00].

Very rapidly, similar plans were also implemented in other countries, with similar priorities to those of the NNI. Accompanied by important financial support, they contributed to induce a strong inclination in science toward all types of nanotechnology. The growth in loans, granted by diverse states or regions of the world, shows rather brilliantly the growing interest in their development (Figure 1.4).

Just as significantly, the increase in scientific publications within this sector also demonstrates a generalized tendency toward the study of nanosciences; quite remarkably, China has now overtaken the United States in terms of publications (Figure 1.5).

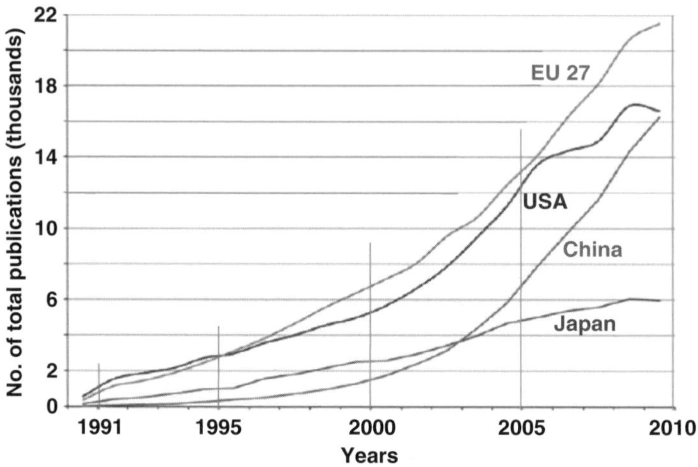

Figure 1.5. *Evolution of the number of total publications within the field of nanotechnology between 1990 and 2009, calculated by studying the keywords in the SCI database (Science Citation Index, Thomson Reuters) (adapted from Roco [ROC 11])*

At the European level, the overall activity in the field of nanotechnology is equally as great. In 2010, it surpassed the United States, China and Japan in terms of total publications, with a distribution according to the 27 members (countries) in the EU, approximately proportional to the size of populations of each one.

Economically, this is a huge market; in 2010, it was given a value of approximately $250 billion.

In terms of research activities, there were 400,000 researchers in 2008 (150,000 of whom are in the United States) within the field of nanotechnology, a number that is constantly increasing and is expected to reach two million (800,000 of whom are in the United States) by 2015 [ROC 11].

In France, since 2005, there has been a major reorganization in research, with the purpose of encouraging the development of nanotechnologies. The MINATEC (Mini and Nano Technologies) center, inaugurated in 2006 at the scientific polygon in Grenoble, currently has more than 2,000 researchers and includes the most significant center of research in Europe dedicated to nanotechnologies; the ultimate aim being the collaboration of international research projects with industry. Academic research was also restructured under the auspices of the French National Centre of Scientific Research (CNRS) with the creation of six regional centers of expertise from 2006 onward (C'Nano) joining together more than 6,000 researchers in the nanosciences and with inclination of favoring exchanges and collaborations between laboratories.

Undoubtedly, society can expect beneficial outcomes to be manifested, especially in the field of medicine as we will see later on; the introduction of nanotechnologies represents a major event in the 21st Century, the same as seen with computing at the end of the 20th Century.

1.4. Bibliography

[CLI 00] CLINTON B., Clinton's Address to Caltech on Science and Technology, 21 January 2000, available at http://marcomm.caltech.edu/events/presidential_speech/.

[FEY 59] FEYNMAN R.P., "There's plenty of Room at the Bottom'', Annual Meeting of the American Physical Society at 'Caltech', December 1959, available at http://www.zyvex.com/nanotech/feynman.html, http://calteches.library.caltech.edu/47/2/1960Bottom.pdf (original).

[INT 11] Intel 22 nm-technology, 2011, available at http://www.intel.com/content/www/us/en/silicon-innovations/intel-22nm-technology.html.

[KIL 00] KILBY J.S., "Turning potential into realities: the invention of the integrated circuit", *Nobel Lecture*, December, 2000, http://www.nobelprize.org/nobel_prizes/physics/laureates/2000/kilby-lecture.pdf.

[NNI 11] NNI Program Components Areas 2001–2012, available at http://www.nano.gov/nnipca.

[ROC 11] ROCO M.C., "The long view of nanotechnology development: the National Nanotechnology Initiative at 10 years", *Journal of Nanoparticle Research*, vol. 13, pp. 427–445, 2011.

[SHO 56] SHOCKLEY W., Transistor technology evokes new physics, Nobel lecture, December, 1956, http://www.nobelprize.org/nobel_prizes/physics/laureates/1956/shockley-lecture.pdf.

[STR 95] STRYER L., "DNA and RNA: molecules of heredity", *Biochemistry*, Chapter 4, 4th edition, Freeman, New York, NY, pp. 87, 1995.

[TAN 74] TANIGUCHI N., "On the basic concept of nano-technology", *Proceedings of the International Conference on Production Engineering*, Part II, Japan Society of Precision Engineering, Tokyo, 1974.

Chapter 2

The Rapid Expansion of Nanotechnology: New Ways of Observing the Infinitesimal and the Discovery of Carbonaceous Nanomaterials with Unusual Properties

Since the 1970s, a large number of scientists have been developing research devoted to the study of surface reactions. Typically, fundamental studies on chemophysical phenomena have been carried out at the atomic and molecular level, localized near the surface (at a region of thickness of several nanometers) and hence this is already within the field of nanosciences.

The real rapid expansion of nanotechnology, however, did not appear until much later, occurring approximately at the beginning of the 1990s, as the publication growth curves show in Figure 1.5, Chapter 1.

There are many reasons for this. Methods for observing the "infinitesimal" improved rather considerably: electron microscopes reached a degree of accuracy that has since allowed the vacuum imaging of surfaces of materials to be done at the atomic level. Moreover, new near-field scanning microscopy techniques, invented from the 1980s onward, became everyday tools, also allowing surfaces to be analyzed up to the atomic level, but this time under environmental conditions, without having to resort to high vacuums, which is the case for electron microscopy.

The expansion of a multitude of diverse types of nanomaterials (NPs, nanowires (NWs) or nanotubes (NTs)), at the same time as the discovery of new carbon nanomaterials with very unusual properties, opened up a huge field of study. This prompted an undeniable enthusiasm toward these new fields of research, which proved to be the most promising in terms of potential innovations.

The domain of the infinitesimal can in effect escape the classical laws of physics; nanomaterials exhibit new properties governed predominantly by the wave theory of quantum mechanics.

This revolutionary theory, whose foundations were established right at the beginning of the 20th Century by renowned physicists (Bohr, Planck, Einstein, Heisenberg, Schrödinger, Dirac, de Broglie, all Nobel Prize winners in physics) brought about great change to the classical laws of physics and chemistry.

This is a theory based on assumptions that do not have anything in common with how we generally perceive such phenomena.

Thus it predicts that it is not possible to simultaneously determine the speed and position of a moving particle, that energy exchanges between electromagnetic radiation (light, for example) and matter are made in a discontinuous fashion, that a wave function, which is associated with each particle, verifies a particular mathematical relationship (Schrödinger's equation), and that this wave function, which replaces the concept of uncertainty and probability with the determinism of classical physics, allows phenomena which occur at the microscopic level to be predicted.

It is this theory that allowed science to make huge progress in the understanding of various phenomena, such as the propagation of electrons and photons in materials, as well as the understanding of molecules and atoms. From that moment, it became an expected tool in the study of nanomaterials as it allowed a better understanding of their properties[1].

1 For an introduction to the concepts of modern physics, see Van Zeghbroeck's lesson accessible via the Internet [ZEG 11]) as well as the book *An introduction to Nanoscience and Nanotechnology* by Alain Nouailhat, Wiley, 2008.

2.1. Improving tools for observing the infinitesimal

Undoubtedly, the different electron microscopy techniques, invented during the 20th Century, have been decisive elements in the progressive understanding of nanomaterials. These techniques are still being perfected; observing objects at the atomic level has become perfectly accessible, especially after the recent invention of near-field microscopes.

Many instruments for analyzing surfaces are currently available on the market; some, however, have very specific properties, whereas others have a more general use. Among the latter, transmission electron microscopy or scanning electron microscopy and near-field microscopes form the basis of the equipment in any specialist laboratory studying nanomaterials and their applications.

2.1.1. *Transmission electron microscopes*

The idea of using electrons to build a microscope appeared in the 1920s in Berlin. Two engineering physicists, Knoll and his student Ruska, specialists in the study of cathode tubes, set themselves the objective of improving the quality of the light spot resulting from the impact of electrons on a fluorescent screen.

This requires a significant know-how in optical electronics, especially a good command over the propagation of electron beams in a vacuum and how they come together at a point on the screen, resulting from electrical and magnetic fields[2] strategically located on their pathway.

On the basis of the extreme small size of electrons and on the workings of optical microscopes, they envisaged building a device in which the beam of electrons (equivalent to rays of light) after passing through a very thin

2 The electric field E is an important physical parameter induced by an electric charge q. It is defined at a point M in space by a vector, that is to say a dimension characterized by a direction (the axis defined by the charge and point M) and intensity proportional to q/R^2 (R being the distance separating point M and charge q). In the case of a capacitor (two metal plates separated by a distance e between which a potential difference (voltage) V is applied), the electric field is proportional to V/e. One of the main properties of an electric field is that it exerts a force on every electric particle found within this field, and as a result, accelerates them, which translates as a drift velocity. (In the case of very high voltage electron microscopes, the speeds of electrons after applying strong electric fields are very fast and can approach the speed of light.)

material sample (to prevent electrons from being absorbed into the totality) would be focused onto a fluorescent screen in order to produce a larger image of the object.

The initial results presented in 1931 by Knoll were, in fact, modest, with magnifications reaching only 15 times; so, the authors did not dare to design their device as a prototype of an electron microscope, but rather as a contribution to the development of geometrical charged-particle optics [RUS 86].

This was definitely the right choice. Also, after coming across the completely new theory of Louis de Broglie, which assigns a wavelength λ inversely proportional to its speed v ($\lambda = h/m \cdot v$) to any moving particle[3], Ruska and Knoll regained confidence in the prospects for electron microscope application and showed that they were going to greatly surpass optical microscopes.

Indeed, the resolving power (that is to say the minimum distance between two points at which they are still distinguishable from one another) is usually about half the wavelength, that is to say for an optical microscope it is absolutely 200 nm, in the best of cases.

By considering that the optical laws of light waves can also be applied to de Broglie wave-particles, it becomes possible, by subjecting electrons to high accelerations (several hundreds of thousands of volts), to give them much faster speeds (of the same order of magnitude as that of light) and hence to associate them with extremely small wavelengths, which will allow a theoretical resolving power in the angstrom range to be attained, that is to say 1,000 times better than the best optical microscope.

Ruska obtained encouraging results fairly rapidly. In 1933, in collaboration with Siemens, he built a microscope with an acceleration voltage of electrons of approximately 70,000 V (70 kV), which showed much better features compared to the optical microscope (magnifications of approximately 10,000 and the projected image being 1 mm for an object of 0.1 μm).

Ruska's whole scientific career was dedicated to this technique, first as an engineer at Siemens from 1937 to 1955, and then as the director of the

---

3 λ is the wavelength associated with the particle, h is a physical constant of quantum mechanics known as the Planck's constant, m is the mass of the electron and v is its speed.

Center for Electron Microscopy at Fritz Haber Institute in Berlin from 1955 to 1972. He also tried to resolve several problems, due in particular to the deterioration of specimens from the impact of electrons, the stabilizing of very high voltages to acquire high-resolution images, and also the application of the technique in biology.

His work was praised in 1986 by the Nobel committee, awarding him the Nobel Prize, shared with two young engineers at IBM (Zürich), who in 1980 had invented another type of microscope, the tunnel-effect microscope.

Considering the huge progress made in technology, the performances currently attained are exceptional. Over the last few years, commercial devices have been operating with voltages often reaching up to 300 kV; this allows magnifications of 300 to 350,000 times to be obtained, which means that an object with a size of 10 nm will be transformed into an image of 3 mm.

These devices can also analyze x-rays (produced after the interaction of electrons with matter), which allow the nature of atoms of the surface under analysis to be identified.

For illustration purposes, Figure 2.1 represents a routine transmission electron microscope (TEM) installation, put on the market in 2007 by the Japanese manufacturer JEOL.

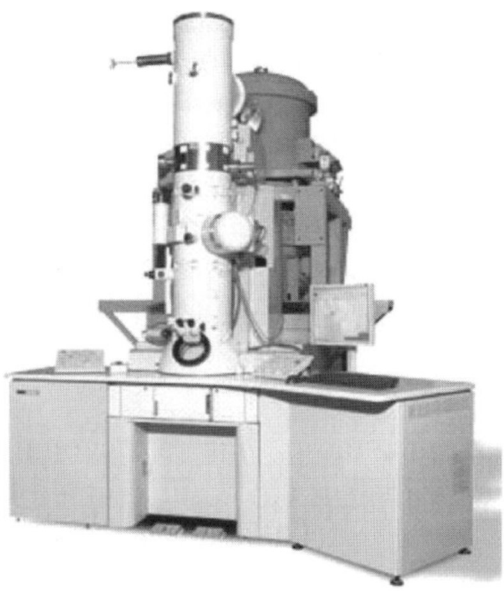

Figure 2.1. *Example of the latest TEM installation, model JEM-3100F, JEOL, 300 kV, 0.17 nm resolution*

Apart from routine commercial devices produced by different manufacturers (JEOL, Leitz, Philips, etc.), transmission electron microscopy has always been the subject of research into high resolutions; also, specialized centers for microscopy have built devices allowing resolutions in the angstrom range to be obtained.

This is the case for National Center for Electron Microscopy, Berkeley, California (NCEM), which, since 2008, has developed a TEM with a very high resolution. This installation allows separation limits of 50 picometer (0.05 nm) to be attained; so now, even the smallest of atoms can be visualized.

The project TEAM, which stands for Transmission Electron Aberration-corrected Microscopy, is a research center for electron microscopy, which has been welcoming international teams working on scientific projects for several years.

Images obtained in this detail are full of information. In particular, they allow us to perfectly visualize the surface of a crystalline material, characterized by regular arrangements of atoms.

The example of a "nanobridge", made up of several single layers of gold atoms between two crystalline gold beads, nicely illustrates the huge potential this technique has for observation. The experiment was carried out by focusing the beam of electrons at different depths below the surface (1.8 nm in Figure 2.2(a) and 7.7 nm in Figure 2.2(b)). In Figure 2.2(b), note how the gold atoms have alignments with distinct orientations defining areas that correspond to weak amplitude "terraces" and "steps" (Figure 2.2).

2.1.2. *Scanning electron microscopes*

The principle of scanning electron microscopy (SEM) is quite different to that of TEM.

Unlike the TEM, where the image is obtained after passing through the sample, in the case of the SEM, the beam of electrons (reduced to only a few nanometers in diameter) is reflected by the specimen.

It interacts with the specimen at a very shallow depth by producing, on the one hand, an emission of low-energy secondary electrons as a result of the ionization of atoms, and on the other hand, a backscatter of part of the incident beam (electrons reflected with a higher energy).

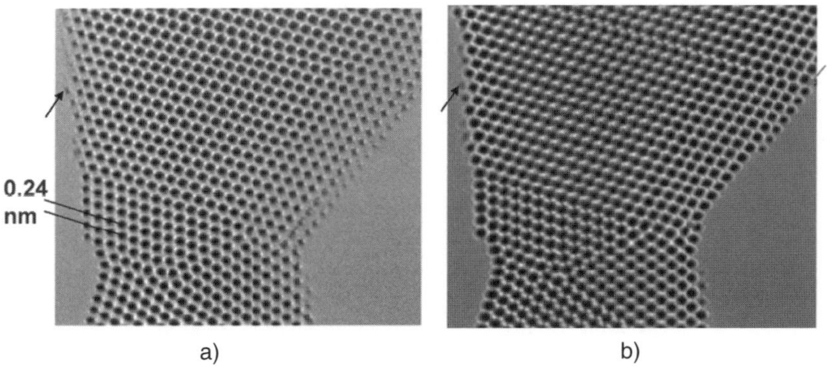

Figure 2.2. *High resolution TEM image of several single layers of gold atoms sandwiched between two thick metallic beads. NCEM Image, TEAM 0.5. We observe images of different orientations of atomic planes corresponding to monocrystalline domains. a) Image obtained by focusing the beam of electrons 1.8 nm beneath the surface. The top left arrow represents a column of two gold atoms. b) Image obtained at a depth of 7.7 nm. The top left arrow corresponds to a single gold atom. The two other arrows on the right represent an alignment of columns of two gold atoms (adapted from Kisielowski et al. [KIS 08])*

The image of the sample is obtained by displacing the incident beam on the surface scanned line-by-line and by recording the intensity of electrons emitted (secondary or backscattered), the best resolution of images being obtained from secondary electrons (Figure 2.3).

Moreover, similar to the case of the TEM, the interaction of the incident beam with the sample is not only restricted to the emission of electrons but also produces an emission of x-rays, which can also be analyzed and used to identify atoms on the surface.

The first devices were designed in Germany by the Knoll group in 1933; they were then modified and improved by von Ardenne in Berlin, and in particular by Oatley at the Cavendish Laboratory in Cambridge.

It was not until the 1960s that the first SEM was put on the market by a British group, Cambridge Scientific Instruments.

The possibility of an increase in size is not as important as it is with the TEM. However, the embossed effect is much greater, being one of the major advantages of the SEM, particularly with its important applications in biology (Figure 2.4).

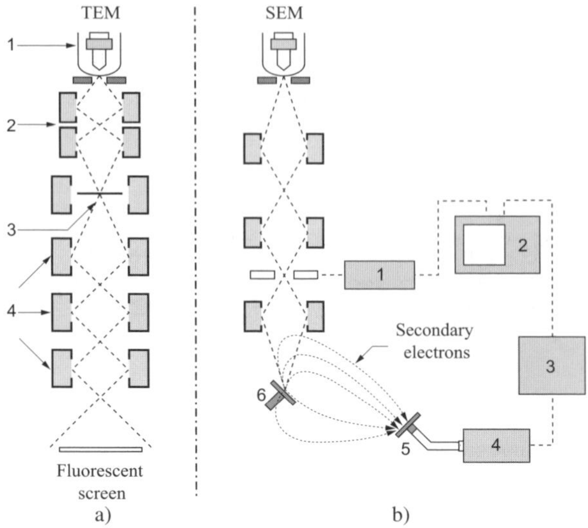

Figure 2.3. *Diagrams of TEM and SEM. a) The electrons are emitted via an "electron gun" 1 (negatively polarized heated filament), accelerated by an anode (positive electrode) and then focused with electromagnetic lenses 2 onto the specimen 3. After passing through the sample (an ultrathin layer) the beam of electrons is increased and projected onto a fluorescent screen. b) In the case of the SEM, the beam of electrons is controlled with the help of a scanning unit 1 ultimately displacing the beam of electrons on the surface of the specimen line by line 6. The flow of secondary electrons emitted off the surface is collected at 5, and then amplified by an electron photomultiplier 4. The information received at 4 is transformed into a potential signal at 3, the two signals (position and intensity of the electron flows) are sent to a screen to be visualized 2 giving an image of the specimen (adapted from J. J. Bozzola [BOZ 99])*

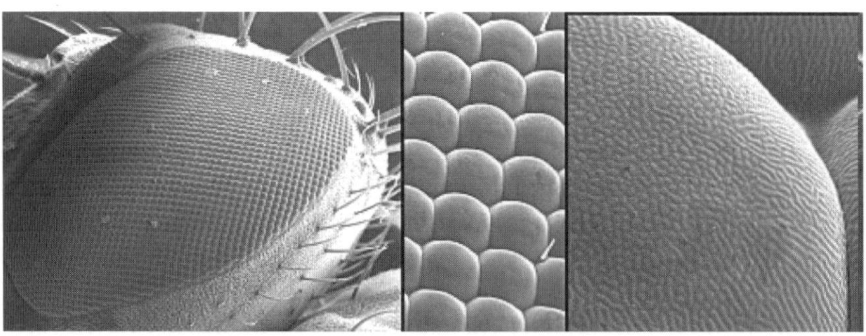

Figure 2.4. *An SEM image of a fly's eye seen at different magnifications (10, 100, 10,000). Electron microscope image courtesy of the Carnot Institute of Bourgogne (http://science-for-everyone.over-blog.com/article-le-meb-microscope-electronique-a-balayage-49566732.html)*

The voltages are lower in the SEM than in the TEM; the magnifications vary from 10 to 500,000 with a maximum resolution of approximately 2 nm.

Furthermore, thanks to recent improvements in techniques, it is also possible to obtain images of objects placed in a moist environment or subjected to weak voltages, which increases the possibilities of SEM imaging of objects that would be unstable under a high vacuum. The image of water droplets in Figure 2.5 nicely illustrates this concept.

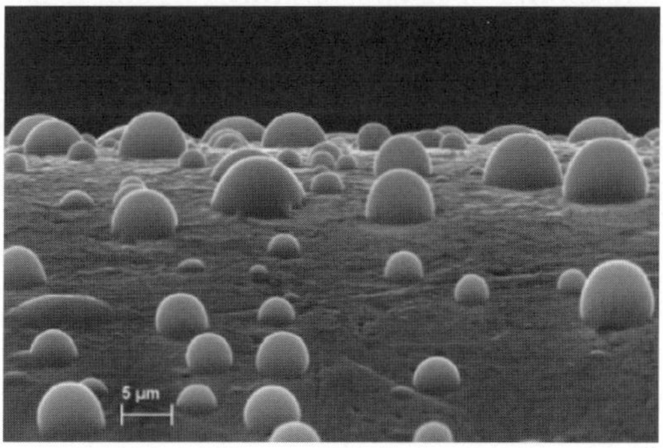

Figure 2.5. *An SEM image of water droplets condensed onto a wire, cooled down to 0.1 °C, under a partial pressure (in Pascals) of 700 Pa (approximately 7/1,000 of atmospheric pressure) and under a 25 kV acceleration voltage of electrons (according to Kugler and Bean [KUG 09])*

2.1.3. *Near-field microscopes*

These completely differ from the electron microscopes that we have previously described, and combine a variety of techniques, which allow, without any geometric magnification, to directly analyze the surfaces at the nanometer scale.

The first of them, the tunnel-effect microscope, was invented in 1980 by Rohrer and Binnig, two engineers at IBM in Zurich. It allows, by measuring the "tunneling current" between a conducting tip and a surface, an image of the object's surface to be reconstructed at the atomic scale.

In 1986, Binnig, Ouate and Gerber developed the atomic force microscope, based on measuring the very small distance of forces of interaction between two objects, thus increasing the possibilities of analyzing insulating objects.

These two techniques revolutionized the study of nano-objects. Thanks to them, it is considered that decisive progress was made in the development of nanoscience because of the multitude of applications to which they have provided.

The considerable impact and unanimous and immediate incorporation into the whole scientific community explains how Binnig and Rohrer won the Nobel Prize so quickly in 1986; they shared it with Ruska in the same year. The latter, as we can see, had to wait for more than 50 years before being rewarded.

2.1.3.1. *The tunnel-effect microscope (STM or scanning tunneling microscopy)*

It operates using the concepts of quantum mechanics, precisely those that directly apply to nanosized objects.

This device involves a very thin conducting tip (only a few microns in diameter), which is moved very close to the surface; a small voltage is applied between this gap (Figure 2.6).

When the tip is too far from the surface, no current flows between them; this is normal. However, when this distance becomes very small (a few fractions of a nanometer), a very weak current is produced, almost localized between the extremity of the tip and the nearest atoms of the metallic surface. This flow of electrons is the tunneling current, predicted by the wave theory of quantum mechanics.

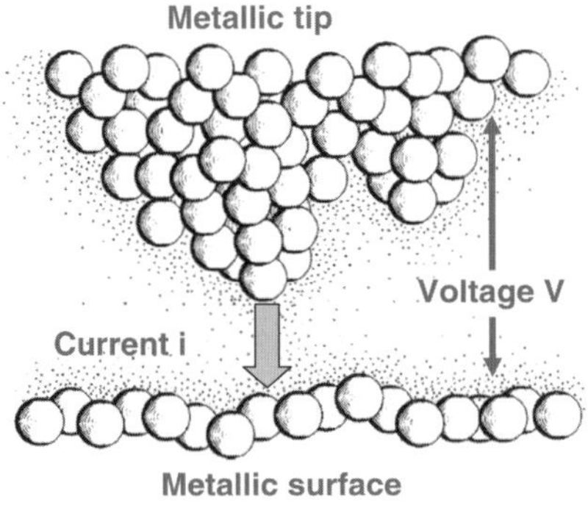

Figure 2.6. *Tunneling current i between the extremity of the metallic tip (shown as an aggregate of clustered atoms) and a metallic surface (represented by its atoms on the surface) between which a voltage V of approximately 1V is applied (adapted from Rohrer and Binnig [ROH 86])*

This current decreases very rapidly as the tip is moved further away from the surface according to an exponential law. (More precisely, the intensity *i* of the current is proportional to $e^{-k \cdot d}$, where *e* represents the exponential function, *k* a constant and *d* the separation distance between the tip and the nearest atoms of the surface.)

An increase or a decrease in the distance by a few angstroms leads to a considerable variation in the tunneling current, which allows the ‘‘tip-surface’’ separation distance to be correlated extremely accurately with the intensity of the tunneling current.

Therefore, it is thanks to this current–distance relationship that it has become possible, by slowly displacing the tip above the surface, to “photograph” the relief by measuring the current.

Therefore, this implies, on the one hand, a software allowing the transformation of the tunneling current to be transformed into level differences of the surface and, on the other hand, an extremely precise mechanical control of the tip in three directions *x*, *y* and *z* at the angstrom scale.

Using piezoelectric sensors (a material that is deformed at low amplitudes under the effect of a voltage) allows these displacement conditions to be accurately obtained. Note that, unlike electron microscopes, where the image is obtained directly from the impact of electrons on a screen, the STM image is indirectly deduced from the tunneling current after a mathematical transformation.

Rohrer and Binnig were accredited for having designed and built a highly complex electromechanical device, whose schematic diagram outlining the principles of operation is represented in Figure 2.7.

The possibilities of using this technique are enormous.

It allows not only observations at the atomic scale to be made but also the manipulation of atoms, and therefore to create objects of atomic or molecular dimensions on the surface.

A team at IBM was the first to show these possibilities in the 1990s by publishing the famous “quantum corral” image in which iron atoms, manipulated with the microscopic tip, are deposited onto a copper crystal in a circle (Figure 2.8).

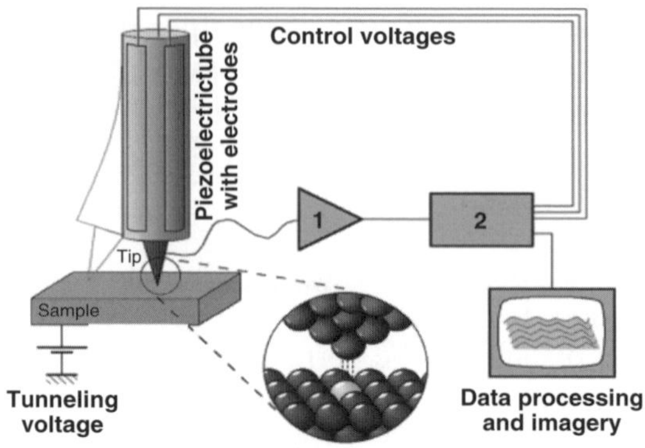

Figure 2.7. *Diagram showing the principle of the tunnel-effect microscope. The piezoelectric tube is a commercial device made up of a piezoelectric material and electrodes, which under the effect of a voltage, cause the material to deform, which translates as displacements of the tip into three directions. The tunneling current is amplified (current amplifier 1) and led toward a control unit (2) whose function is to act upon the piezoelectric tube in a way that brings the tunneling current back to its initial value, thus keeping the tip-surface distance constant. The voltages that control the displacements and the current are then subjected to computer processing of data, allowing the height differences in the surface to be directly recorded according to the displacements in the directions x and y (adapted from Wikipedia)*

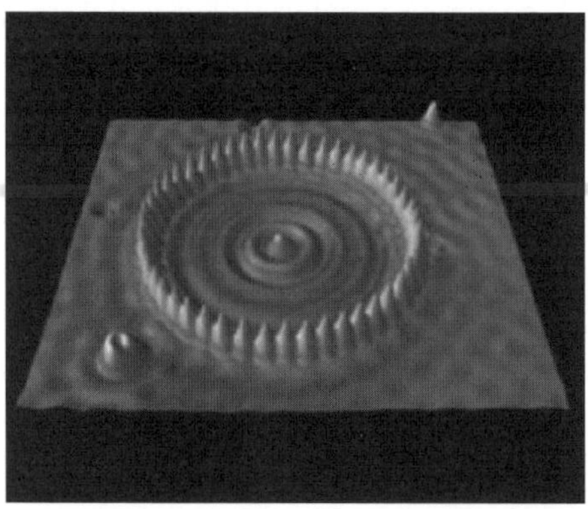

Figure 2.8. *STM image of iron atoms deposited in a circle onto a crystalline surface of copper ("quantum corral"). The image is obtained by manipulating iron atoms in a vacuum at a very low temperature, IBM Corporation, Almaden, 1993 http://www.almaden.ibm.com/vis/stm/. For a color version of this figure see www.iste.co.uk/lacaze/nano.zip*

Thanks to this technique, other equally spectacular images show that the last stages of a chemical reaction can be visualized at the molecular level too.

The Ullman reaction, well known by organic chemists as the process in which iodobenzene (C_6H_5I) reacts at high temperatures (350°C) and in the presence of copper to produce phenylbenzene ($C_{12}H_{10}$ or H_5C_6–C_6H_5) and copper iodide (CuI_2) has been revisited by STM techniques.

Hla *et al.* [HLA 00] also succeeded, via the tip of the STM microscope, to split a molecule of C_6H_5I into two fragments, C_6H_5 and I, with the two groups of C_6H_5 recombining into a new product, phenylbenzene.

This manipulation, which requires great expertise, must be carried out in a vacuum and at extremely low temperatures (a low temperature is needed to immobilize the molecules on the surface of the copper).

The different stages of the reaction (at the molecular level) are represented in Figure 2.9. Figure 2.9(a) shows two molecules of C_6H_5I on a copper strip (two bulges in the gray upper part of the figure or yellow in the on line colored figure) above the black (red in the on line figure). The tip comes into contact with the molecule, and is subjected to an electric impulse, which causes the bond between I and C_6H_5 to break.

Figure 2.9(b) shows an enlarged bulge, which corresponds to the effective dissociation of two fragments, I and C_6H_5, of the molecule. The same operation is carried out on the second molecule (Figure 2.9(c)).

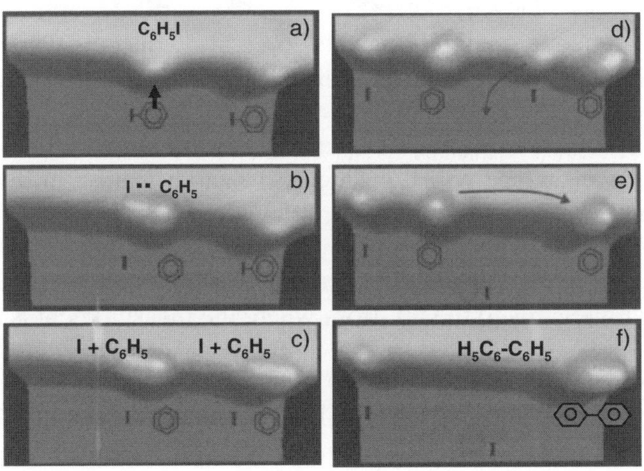

Figure 2.9. *STM images, step-by-step of the molecular reaction 2 C_6H_5I + 2 Cu = $C_{12}H_{10}$ + 2 CuI (adapted according to Hla et al., [HLA 00]). For a color version of this figure see www.iste.co.uk/lacaze/nano.zip*

The iodine atoms (I) are separated with the help of the tip (Figure 2.9(d)); the bonding of the two phenyl groups C_6H_5 is always carried out with the tip (Figure 2.9(e)), to ultimately synthesize, after another electrical impulse, a molecule of phenylbenzene $C_{12}H_{10}$ (Figure 2.9(f)).

These images illustrate the extraordinary possibilities of Scanning Tunneling Microscopy and nicely show that, thanks to the possibility of manipulating atoms or molecules, great prospects are offered to specialist researchers in nanosciences.

2.1.3.2. *Atomic force microscopy*

The principle of atomic force microscopy (AFM) is completely different from STM and, in this case, it is based on measuring the forces interacting between a tip and a surface, which only appear when the tip and the surface are extremely close to one another.

These forces of interaction are very weak, they are measured in piconewtons[4] and their intensity decreases rapidly with increasing distance.

The tip is mounted onto an elastic cantilever, whose point of equilibrium (corresponding to the equality between the force of interaction and the restoring force of the cantilever) is controlled with the help of an optoelectronic feedback device.

As a consequence, maintaining this force at a constant value once again involves fixing the distance between the tip and the surface.

Obviously, by very slowly displacing the tip above the surface, any unleveling of the surface will translate into a modification in the interacting force and therefore as a displacement of the cantilever.

This brings it back to its equilibrium position by a relative displacement of the surface with regards to the tip, from where the measurement is deduced.

Similar to STM, the displacements x, y and z are controlled to a tenth of a nanometer by piezoelectrics, which allow, in the same way as with STM, to visualize and record the topography of the surface according to displacements in the x–y plane (Figure 2.10).

---

4 One newton is an approximate force which must be exerted to lift a weight of 100 g; the piconewton represents a trillionth of a newton or 10^{-12} N.

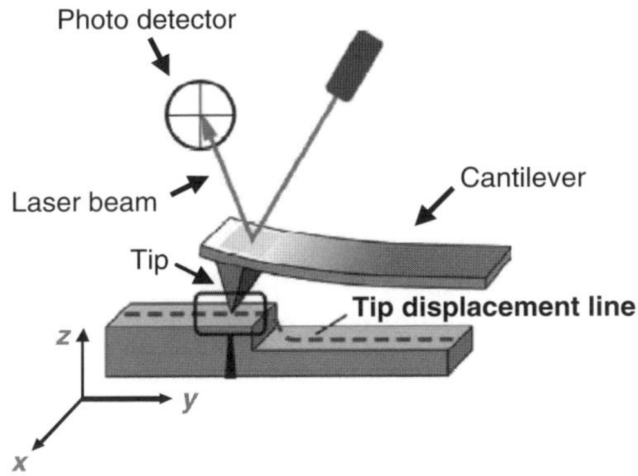

Figure 2.10. *Diagram showing the principle of AFM. The cantilever is a flexible blade kept at a fixed distance from the surface with the help of an optoelectronic feedback device (adapted from a lecture by N. Frangis, http://www.mansic.eu/documents/PAM1/Frangis.pdf)*

The advantage of this technique with regards to STM is that it can be applied indifferently to insulating or conducting materials. It is used a lot in biology and allows many macromolecules to be visualized on a surface such as strands of DNA[5] (Figure 2.11).

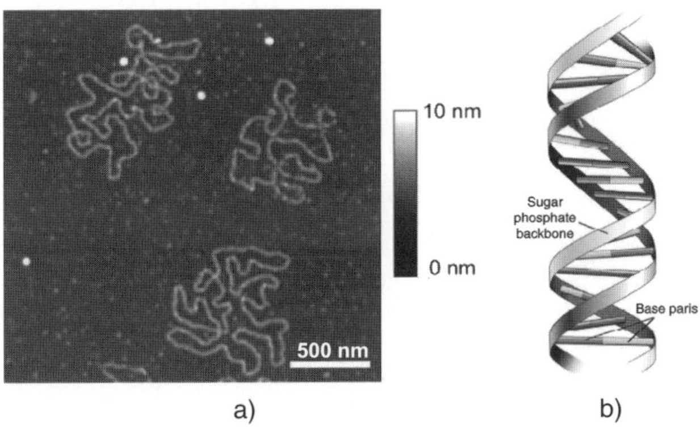

Figure 2.11. *a) Image of simple strands of DNA deposited onto mica after DNA is processed by a chemical compound. This is to make the DNA strands interact with the mica surface more easily; mica being chosen because of its flatness. The color key shows an average thickness of the DNA strands of approximately 2.5 nm (according to Hamon et al. [HAM 07]). b) Diagrammatic representation of a double strand of DNA resulting from the association of two complementary strands, linked via pairs of purine bases (adenine-thymine and cytosine-guanine or A-T and C-G). For a color version of this figure see www.iste.co.uk/lacaze/nano.zip*

5 See Box 3.3, DNA (deoxyribonucleic acid).

2.2. The discovery of new carbonaceous nanomaterials

Carbon is the next most abundant element in the universe after oxygen and helium (the majority of helium being found in stars).

Undoubtedly, carbon is the most abundant element in the biological domain; it forms the largest component of organic compounds, whether from a natural biological or from a synthetic origin. This omnipresence of carbon in the majority of organic compounds is due to the unusual structure of the carbon atom: it has a small number of electrons surrounding the nucleus.

This is also due to its middle position in the second row of Mendeleev's periodic table, which makes it one of the only known elements to form unlimited bonds with itself (formation of macromolecules and polymers characterized by –(C–C)–*n* chains, and also capable of forming stable bonds with most of the other elements[6].

Since the 1980s, the discovery of a new family of carbonaceous nanomaterials similar to graphite (fullerenes, NTs of carbon and, most recently, graphene) has undoubtedly generated a new interest in the study of nanotechnology, explained solely by the innovative and countless possibilities resulting from it.

Each of these carbonaceous nanomaterials, whose structures are similar to that of graphite, differs in dimension.

The fullerenes (very tiny spherical NPs) are adimensional or 0D, NTs (similar to nanowires) can be considered as one-dimensional (1D) whereas graphene, due to the thickness of its single layer of carbon atoms, is typically a two-dimensional (2D) material. This concept of different dimensions is important, as it affects the groupings of electrons and, therefore, the physical and chemical properties of the materials.

---

6 Mendeleev was a Russian chemist who became famous for having proposed a basic periodic table of elements in 1869, in which different known elements were arranged in order of increasing mass and laid out in different rows and columns. This classification is interesting as elements with similar chemical properties show up in the same column. However, this table has since been completed and modified completely by no longer arranging each element according to increasing atomic mass but according to increasing atomic number; that is by the increasing number of electrons arranged around each nucleus. Carbon with atomic number 6 (six electrons surrounding the nucleus) occupies the very middle position in the second row of the table (which is composed of the elements lithium, beryllium, boron, carbon, nitrogen, oxygen, fluorine and neon).

In order to better comprehend the effect the electronic configuration has on the properties of these materials, we are going to recap some fundamental principles about the chemical bonds between carbon atoms.

Graphite and diamond, both composed solely of carbon, are two extreme cases. The extraordinary thing is that their properties are completely opposite to one another: diamond is insulating, transparent to light and incredibly hard; graphite is a conductor, a lubricant, black and opaque.

Why are there so many differences in the properties of these materials? The answer is that diamond does not possess any "free" electrons whereas in graphite, which is composed of flat parallel layers of carbon atoms, the electrons can flow freely in the spaces between the layers.

2.2.1. *Some basic concepts relative to the electronic structure of carbon and to the bonding rules between carbon atoms*

To make things simple, theoretical chemistry (derived from quantum mechanics) says that the electrons orbiting the nucleus of an atom are arranged in different energy levels with a very large tendency to occupy particular regions in the atoms, called orbitals[7].

Moreover, when two atoms come into contact to form a chemical bond, it is by the "fusion" of two atomic orbitals into a new "molecular" orbital that the chemical bond will be formed, corresponding to the sharing of two electrons from two separate atoms.

This type of bond is a covalent bond, characterized by a high-energy bond stability, which makes the bond between the two atoms very strong.

2.2.1.1. *The enigma of carbon atoms*

Each atom has its own chemical properties.

In the case of carbon, the electrons (six in total) are distributed over two energy levels; two electrons in the first and four in the second. The four electrons in the second energy level are not bonded to the nucleus as strongly as the two in the first energy level.

---

7 In fact, an atomic orbital is a mathematical function associated with an electron. It defines a spatial region, with a nucleus in the center, and where the likelihood of an electron existing is greatest.

As a result, these four electrons, with more independence than the first two, will have the possibility of interacting with electrons in other atoms to form covalent chemical bonds (this is why they are referred to as valence electrons).

An additional feature of carbon is that through the process of hybridization (the mixing of atomic orbitals), it can have two, three or four of these valence electrons in particular spatial regions of the atom, called hybrid atomic orbitals.

The first type of mixing, applied to four electrons, leads to the formation of four similar "pear-shaped" orbitals, called sp^3 orbitals. The bond angle between each axis is 109.5° (Figure 2.12).

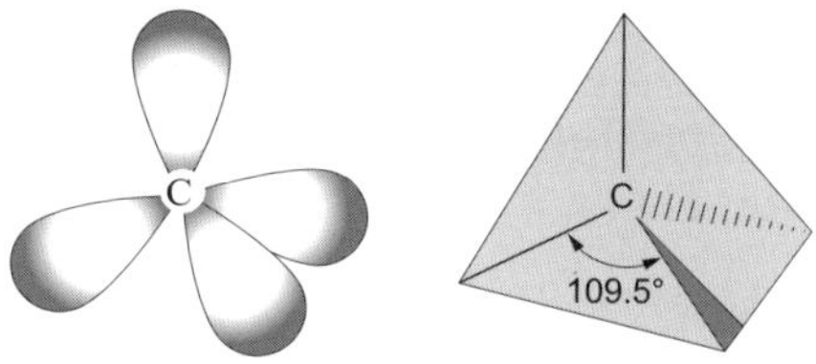

Figure 2.12. *Spatial representation of four sp^3 orbitals of a carbon atom*

This type of structure can be found in most organic compounds (hydrocarbons, polymers, other carbonaceous compounds). They are characterized by linear or branched chains of carbon atoms of variable lengths, combined with hydrogen or other atoms.

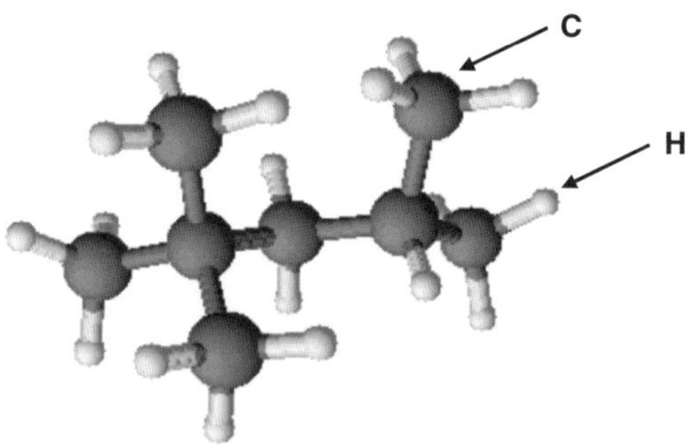

Figure 2.13. *Spatial representation of a molecule of isooctane (eight carbon atoms, 18 hydrogen atoms). All the interatomic bonds have an angle of 109°. The molecule is highly flexible because of the fact that the different atoms, taken two-by-two, can turn freely about their bonds*

Molecules with these types of chains are inert chemicals. This is the case for saturated hydrocarbons, the main components of gasoline (which has a high percentage of octane molecules, such as isooctane which is liquid at room temperature (Figure 2.13)).

Note that for this type of compound, all the valence electrons of carbon are bonded, and each carbon atom is surrounded by at least eight electrons (two electrons being shared in each bond).

This electronic configuration (highly stable) obeys the octet rule, which states that atoms of a low atomic number can only contain, in their outer electron shell, a maximum of eight electrons.

2.2.1.2. *Diamond or the perfect and unique tetrahedral chain of carbon atoms*

It is also from this sp^3 hybridization state of carbon that diamond crystals are formed; each carbon atom in each crystal is bonded to four other carbon atoms, with bond angles identical to those formed by sp^3 orbital axes (Figure 2.14).

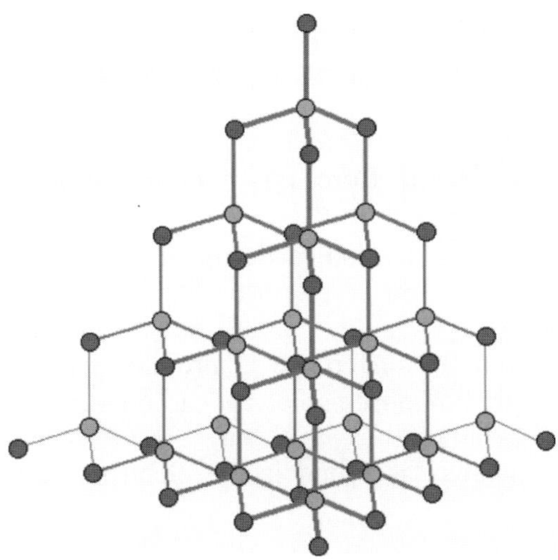

Figure 2.14. *Visual representation of the structure of a diamond crystal. All the bond angles between the carbon atoms are 109°; this is characteristic of an sp^3 hybridization*

Unlike the previous case for hydrocarbons, the structure is extremely rigid. There is no longer any free rotation at the bonds, the carbon atoms are firmly bonded to one another and can no longer move or rotate about the –C–C– bonds.

It is precisely this structure, which is at the root of these extraordinary properties of diamond: hardness, transparency and perfect electrical insulation.

Its hardness is due to the great stability of its covalent bonds, which leads to a strong cohesion of carbon atoms with one another; it is transparent due to the fact that all its electrons are strongly attracted to the carbon nucleus with high-energy stabilities; this prevents light from "dislodging" these electrons from their "fundamental" state. The light will therefore pass through the material without being absorbed at all.

This is also the reason why it is an electrical insulator. The bonding electrons are strongly attracted to the nucleus and are incapable of moving about under the effect of an electric field.

2.2.1.3. *Graphite or the intrusion of π electrons in the assembly of carbon atoms*

This is another type of orbital mixing, better known as sp^2 hybridization. This forms the basis of (all ethylene or benzene) sequences of unsaturated carbon compounds.

Graphite is also formed from this type of structure, with physical properties completely opposite to those of diamond: it is opaque and black, conducts electricity, and it is also made up of layers, giving it lubricating properties.

How do we explain the difference in properties between diamond and graphite, when both are composed solely of carbon atoms?

Once again, the answer is found in the electronic configuration of the carbon atoms.

In the case of graphite, this time it does not involve the sp^3 hybridization, but an sp^2 hybridization.

This means that the four valence electrons are distributed between three identical sp^2 atomic orbitals and one p orbital (denoted by p_z), which has not been used in the hybridization process.

From the geometric point of view, this leads to three sp^2 orbitals with similar shapes as sp^3, but this time, the axes are situated in the same plane and have bond angles of 120°.

That leaves another atomic orbital (denoted p_z), whose axis is perpendicular to the plane formed by the three axes of sp^2 atomic orbitals. This orbital contains one electron and is not involved in the hybridization process. (Figure 2.15(a)).

Just like with diamond and sp^3 orbitals, sp^2 orbitals can fuse with sp^2 orbitals of other carbon atoms, to form flat hexagonal shapes, corresponding to a sheet of graphene.

The question asked is: what happens to the electron situated in the p_z orbital?

The answer is simple: because of the parallelism between the axes of these p_z orbitals and their great proximity, there is a lateral fusion between p_z orbitals, which leads to the total delocalization of these electrons in the structure.

These delocalized electrons in a network of carbon atoms are called π electrons, unlike the electrons making up the covalent bonds between two carbons, called σ electrons.

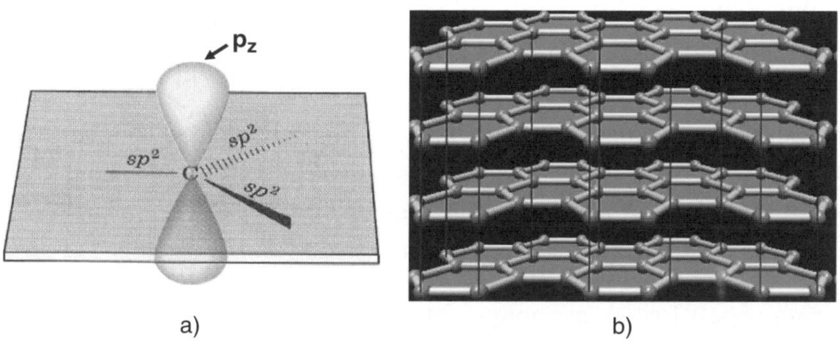

a) b)

Figure 2.15. *a) Diagrammatic representation of sp^2 and p_z orbitals. b) Diagrammatic structure of graphite: the carbon atoms are bonded at the corners of regular hexagons (bond angles of 120°). They form sheets of graphene bonded to one another with weak forces due to the interaction of π electrons*

It is the π electrons, delocalized on both sides of the sheet of graphene, which by interacting with one another allow a strong enough cohesion to be maintained between the sheets, and so the structure of graphite can be formed by stacking thin layers of graphene (Figure 2.15(b)).

This particular structure of graphite makes it an anisotropic material[8].

The presence of delocalized π electrons between the layers generates a very large conductance of electricity in the directions parallel to the graphene sheets and, on the contrary, a weaker conductivity in the direction perpendicular to the sheets.

For the same reason, the π electrons are not attached to any carbon atom in particular; the sheets can each slide over one another in various directions, making it a good lubricant.

Finally, the opacity of graphite and its black opaque color can also be explained by the fact that these same π electrons, bonded to carbon atoms by weak forces, can be "excited" by light (dislodged from their initial state); this causes total absorption of light and hence the black color of graphite.

2.2.2. *The fullerenes or graphite sheets rolled into a ball*

Discovered by Curl, Kroto and Smalley in 1985, they are made up of an assembly of hexagonally and pentagonally arranged carbon atoms with an overall spherical shape; this gives them structures similar to the balls used in soccer, which is why they can also be called buckyballs (Figure 2.16).

This discovery was the result of an unusual observation, after a long discussion between three scientists, Kroto, Curl and Smalley, whose areas of research were very different; because of this, they were in no way predisposed to successful collaborations.

Kroto was a British astrophysicist at the University of Sussex who was interested in the study of interstellar carbonaceous compounds. Curl and Smalley were two American physicists at Rice University, Houston, working on the synthesis of small carbonaceous molecules obtained by laser ablation of graphite (removal of material with a laser beam).

---

8 An anisotropic material has physical properties that depend on direction.

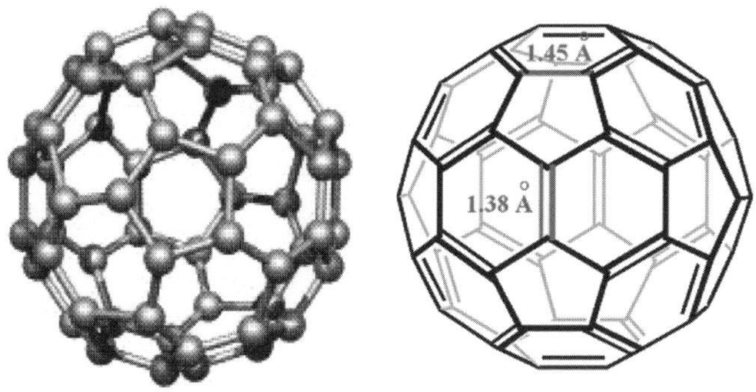

a) b)

Figure 2.16. *Images of a C_{60} molecule. a) A model representing the bonding structure of carbon atoms, b) model formed from an assembly of regular hexagons (benzene nuclei) and pentagons with alternating double and single bonds, with lengths of 1.38 and 1.45 Å respectively. This structure makes this molecule extremely stable. Moreover, the presence of double bonds allows other chemical groups to bond, giving it new properties*

What they had in common was that they had discovered an unknown compound, found in minute quantities, composed solely of carbon complying with compositions of 60 carbon atoms.

Finding a chemical formula for this compound was a daunting task. After being awarded the Nobel Prize, Kroto told rather lightheartedly, how the idea of the structural formula of C_{60} came into his mind [KRO 96].

After having viewed the geometry of the geodesic dome at the American pavilion of the 1967 Universal Exposition in Montreal, due to Buckminster Fuller, the structure of C60 became clear to Kroto; it resembled an assembly of C atoms, similar to that of a geode.

After many discussions and with the help of various chemophysical characterizations, the structure of the material was finally established in 1985. The authors showed that the structure of C_{60} corresponded to an assembly of 60 carbon atoms; 20 regular hexagons and 12 pentagons arranged into a sphere [KRO 85].

The discovery of these materials, named “fullerenes”, in honour of Buckminster Fuller, immediately generated a lot of excitement, and numerous research groups were created to study these materials in order to find diverse applications.

One of their most remarkable properties is that these are “electron wells”, which enables them to be used strategically as materials for transporting electrons in plastic solar devices, as well as in the production of organic light-emitting diodes (OLED) (Figure 2.17).

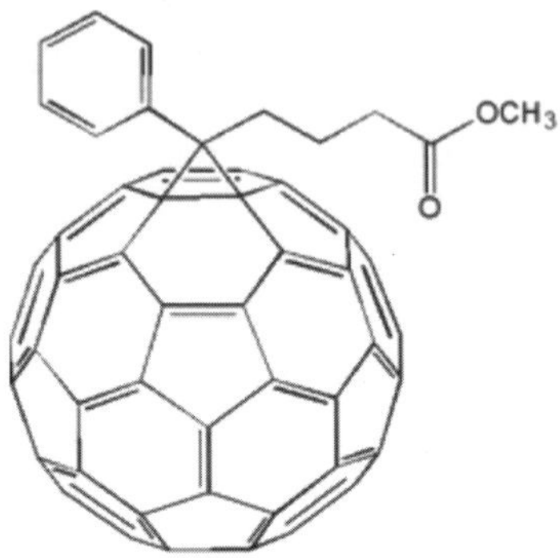

Figure 2.17. *Fullerene derivative (PCBM or [6,6]-phyenyl-C_{61}-butyric acid methyl ester) used as a semiconductor (conduction by electrons) in the creation of organic photovoltaic cells or OLED (see Chapter 6)*

Very quickly, new methods of synthesis were proposed. At the same time, a chemistry of the fullerenes was developing from the elaboration of fullerenes with a variety of chemical groups.

Currently, numerous societies are able to supply large quantities of these nanomaterials on demand, whether functional or not.

Ten years after these important nanomaterials were discovered, Kroto, Curl and Smalley shared the Nobel Prize for Chemistry in 1996.

2.2.3. *Carbon nanotubes: tubes of graphite sheets*

Discovered, several years after the fullerenes, by Lijima in Japan [LIJ 91], carbon nanotubes (CNTs) have since proved to be extremely interesting in terms of their remarkable electronic and mechanical properties.

They appear as tubes only a few nanometers in diameter but with lengths of several microns; the carbon atoms are found at the corners of regular hexagons with a similar geometry to graphite sheets, which can roll up on themselves (Figure 2.18).

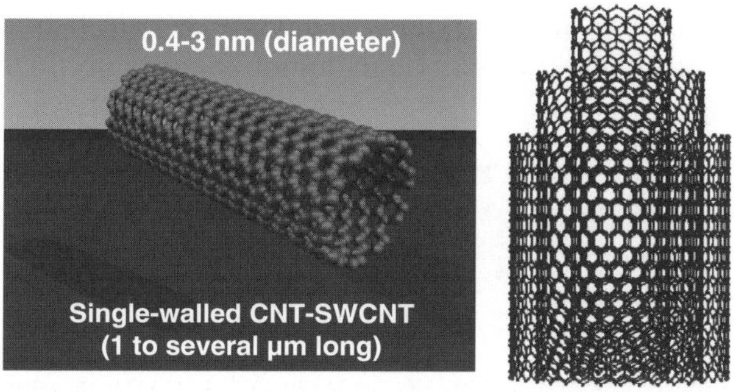

Figure 2.18. *Representation of a single-walled CNT and multiwalled CNT*

Single-walled CNTs (SWCNTs) are formed by rolling a single sheet of graphite into a cylinder.

They can also be found in the form of concentric nanotubes, and in this case they are called multiwalled CNT (MWCNT).

Also discovered in trace amounts, and initially obtained by applying electrical discharges between electrodes of graphite, they are now produced in significant amounts, and just like fullerenes, they have become commercial products that can be easily found.

Their mechanical properties are extraordinary.

They have a tensile strength[9] of 100 to 200 times greater than that of stainless steel, also much more superior than those of other carbonaceous materials such as carbon fibers or Kevlar (Figure 2.19).

This explains why they are used to reinforce plastics, and why they have recently appeared in diverse pieces of sporting equipment: tennis rackets, baseball bats, bicycle frames and forks, etc.

Their electrical properties are also astounding.

9 The tensile strength is the maximum stress a material can withstand while being stretched or pulled before necking.

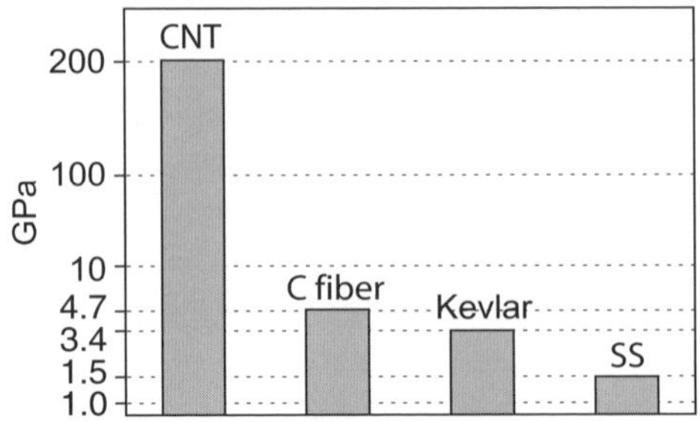

Figure 2.19. *Mechanical performances of CNT compared to carbon fibers (C fiber), Kevlar and stainless steel (SS) (according to Loiseau, ONERA [LOI 01])*

Their conductibility differs depending on whether they are found in the single-walled form (conductor or semiconductor (SC)) or in the multiwalled form (insulator).

A special feature of these carbon nanotubes is that they have highly mobile π electrons (the same found in graphite) according to the axis of the tube (10–100 times greater than that of silicon).

This high mobility corresponds to the electron speed generated when an electrical field is applied. This proposes promising outlooks in the field of nanoelectronics.

Given the progress made in the manufacture of CNTs, the Intel® Corporation predicted that by 2012, CNT-based field-effect transistors would be constructed on a large scale (Figure 2.20), which would lead to electronic components with much greater switching speeds than those of silicon-based components.

Another equally interesting property is how easily CNTs can release electrons from their ends, when they are subjected to an electric field.

Using CNTs as field emitter displays to produce ultra-flat television screens was already proposed in 2005 by Motorola, Applied Nanotech Inc. and Canon, and was presented as one of the possible developments for television screens.

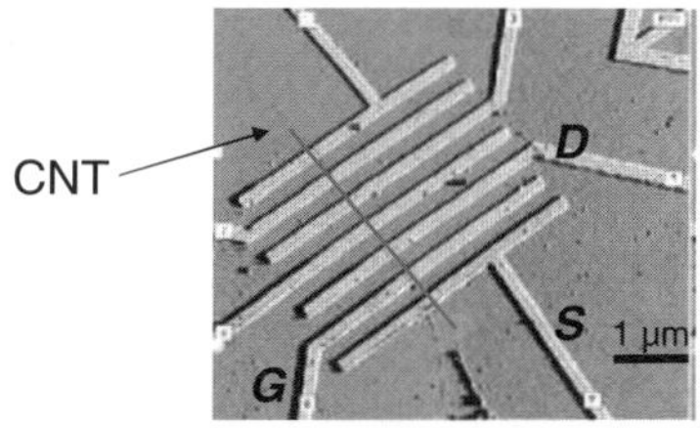

Figure 2.20. *CNT-based field-effect transistor. The CNT is placed at a right angle to two metal electrodes S and D (source and drain) between which a weak voltage V_{SD} is applied. Applying an electric field to CNT via an electrode G (gate), set at a potential V_G and placed at a small distance from the CNT, allows the current flowing through the CNT to be modulated and is collected by the electrodes S and D (adapted from a document, Intel® 2011)*

Similar to fullerenes, we can also contemplate chemically modifying these CNTs and adding chemical groups onto their walls and ends and hence altering their chemical properties, to give new applications.

2.2.4. *Graphene or graphite "sheets"*

This is probably the most recent discovery that has excited the scientific community the most, due to its extraordinary properties. It has even been said that graphene is perhaps the material of the 21st Century.

2.2.4.1. *The identification of graphene*

Initially it was an experimental challenge that simply involved removing, by a process of exfoliation, the sheets of carbon making up graphite, to isolate them on a silicon substrate and manipulate them in order to study their physical properties.

This mechanical procedure, similar to peeling an onion, proved to be reliable enough for the two physicists, Geim and Novoselov of the University of Manchester, to carry out an accurate study on the electrical properties of this new material, named graphene (Figure 2.21).

Remarkably, this sheet of graphene is made up of a single layer of carbon atoms, positioned at the corners of regular hexagons, and it is highly stable, both chemically and mechanically, and can therefore be manipulated and incorporated into measuring devices.

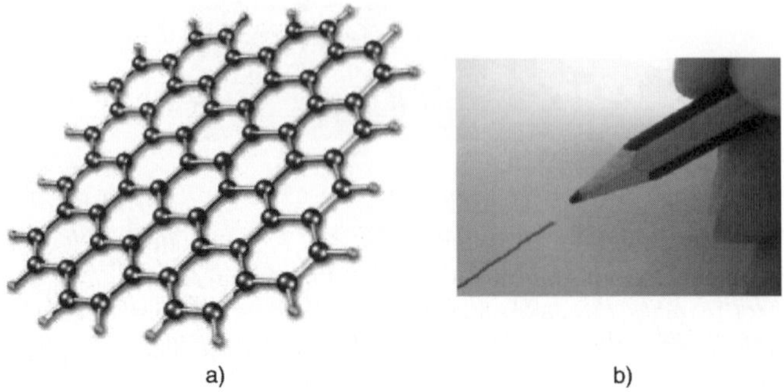

Figure 2.21. *a) Diagrammatic representation of a sheet of graphene. The dark balls represent carbon atoms and the light balls represent hydrogen atoms bonded to the carbon atoms at the edge of the sheet; b) graphene sheets produced by drawing on paper with a pencil*

For physicists, this is an excellent opportunity, because it is the first time that they have a stable 2D material composed of a single layer of atoms at their disposal. They will therefore be able to verify Dirac's theory established in the 1920s, concerning the behavior of delocalized electrons in a 2D network of atoms (in this case, π electrons are involved, the same as those in graphite).

This is precisely what Geim and Novoselov brilliantly did in 2004, which earned them the Nobel Prize in 2010 [NOV 04].

Geim and Novoselov in fact showed great skill in the experimental manipulation of this new material.

The exfoliation method, which involves peeling graphite with scotch tape or through filing by rubbing it against a solid surface is insignificant, except when subsequently needing the identification of graphene "sheets", which at this thickness of only a few angstroms, are completely transparent.

They became visible when the surface was observed by AFM and after depositing them onto a very flat silicon base, oxidized into silica on the surface (Figure 2.22).

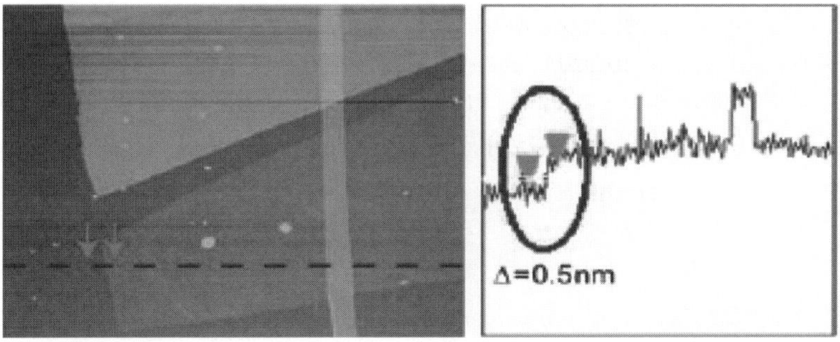

Figure 2.22. *AFM observation of graphene sheets placed on a substrate of silicon dioxide (SiO_2) also known as silica. The dotted line indicates the path followed by the tip and confirms the presence of graphene on the surface with a sharp level difference of 0.5 nm corresponding to the thickness of the sheet (according to Chen et al. [CHE 07])*

2.2.4.2. *Some remarkable electrical properties*

Graphene is completely stable at room temperature, unlike all other 2D materials that degrade when placed under the same environmental conditions. Geim and Novoselov showed that the mobility of electrical charges (displacement speed under the effect of an electric field) is much higher (300 times greater than that observed for silicon).

This property is particularly interesting because it offers completely new perspectives for the design of field-effect transistors, with extremely quick switching speeds.

Such a characteristic has recently been implemented by researchers at IBM. They have in fact designed a field-effect transistor with a graphene sheet that allows electrical signals to be amplified with switching speeds of approximately 100 GHz[10], whereas the same transistors based on Si are only able to obtain, in the best situation, approximately 40 GHz [LIN 10] (Figure 2.23).

The result is even more remarkable than the device designed with technology used to make silicon-based transistors; this could effectively be the first step toward the manufacture of these new types of components at the industrial scale.

10 Gigahertz is a frequency value; 1 GHz is equal to a billion oscillations per second.

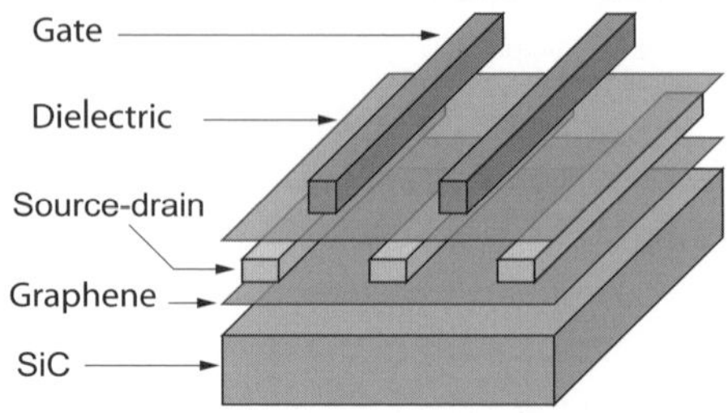

Figure 2.23. *Field-effect transistor with a graphene sheet (according to Lin et al. [LIN 10]). For a color version of this figure see www.iste.co.uk/lacaze/nano.zip*

Samsung, another large manufacturer of electronic materials, has also visualized another application for graphene, by showing that it can be used as a transparent metallic electrode to make plastic television screens; in 2011, they announced that they were manufacturing graphene sheets with dimensions of 1 m^2 (Figure 2.24).

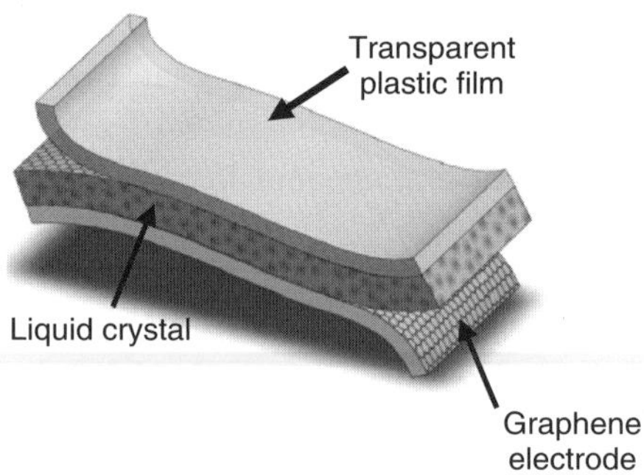

Figure 2.24. *Plastic television screen, based on liquid crystals and transparent graphene electrodes. Using flexible and transparent graphene sheets on both sides of the liquid crystal layer replaces the traditional assembly of glass slides coated with a conductive coating (an alloy: indium tin oxide or ITO) (adapted from a Samsung document, 2011)*

This announcement was made alongside the development of an almost industrial process, for the manufacture of rolls of graphene, by several research groups from Korea, Japan and Singapore. This result is special, in the sense that it is based on the implementation of traditional microelectronic production techniques and involves chemically condensing a mixture of methane and hydrogen on a flexible film of copper in a vacuum in the presence of catalysts [BAE 10].

2.2.4.3. *Remarkable progress: solid, flexible and easily manipulated graphene paper*

Following the initial publications by Geim and Novosolov in 2004, who verified the extraordinary physical properties of graphene, a collection of properties never before found in a material of this kind, graphene instantly became interesting to many researchers throughout the world. They have since tried to develop methods that allow this material to be produced in large quantities.

Currently, it is known how to create a type of graphene in the form of sheets of paper.

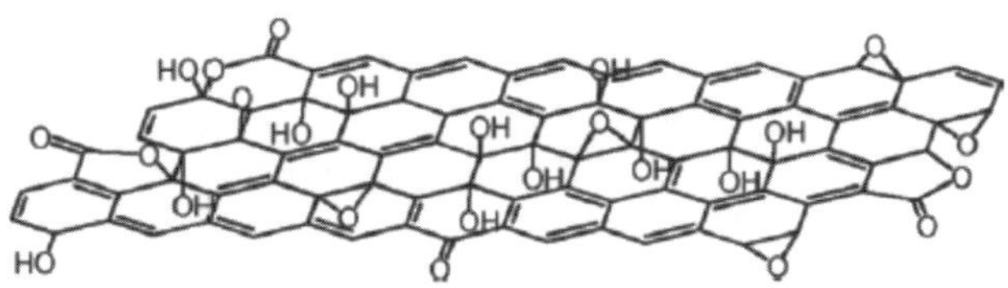

Figure 2.25. *Structure of a sheet of graphene oxide. Note that unlike graphene, double and single bonds no longer alternate in a regular fashion, because of the formation of C=O, C–OH and –C–O–C– bonds. This is also the reason for the electrical conductivity of graphene oxide being very low compared to that of graphene (according to Gao et al. [GAO 09])*

Diagrammatically, the transformation involves oxidizing the graphite powder to form graphite (GO) in the first stage, equivalent to a sheet of graphene with C=O and OH groups bonded (Figure 2.25).

Graphite oxide is then exfoliated, giving sheets of graphene oxide equally dispersed in a solvent; next, the solution is chemically reduced to restore the fragments of graphene.

The vacuum filtration of a suspension of graphene fragments causes them to clump together and form a kind of paper with a metallic, rather homogeneous, appearance, with mechanical properties (resistance and flexibility) just as good as those seen in graphene sheets (Figure 2.26).

The thickness of graphene sheets can be adjusted by using more or less concentrated graphene dispersions. The electrical conductivity of graphene paper is also very good, which has already allowed it to be used in numerous applications concerning the storage and conversion of energy, in biology, in electronics, etc. [BAI 11].

Among the many recent innovations, we can also point out the use of graphene paper in the manufacture of flexible and hyperactive electrodes; this nicely shows the high conductivity of this material.

This very clever procedure involves making a sheet of graphene oxide with a thickness of only a few microns; it is endowed with impressive mechanical properties, and is therefore resistant enough to be manipulated without being damaged.

This sheet of graphene oxide is a weak conductor and is used as a solid base, on which gold NPs can be compiled in a compact fashion. The conductivity of graphene is reestablished by electrolysis, which causes graphene oxide (GO) to transform into graphene (G).

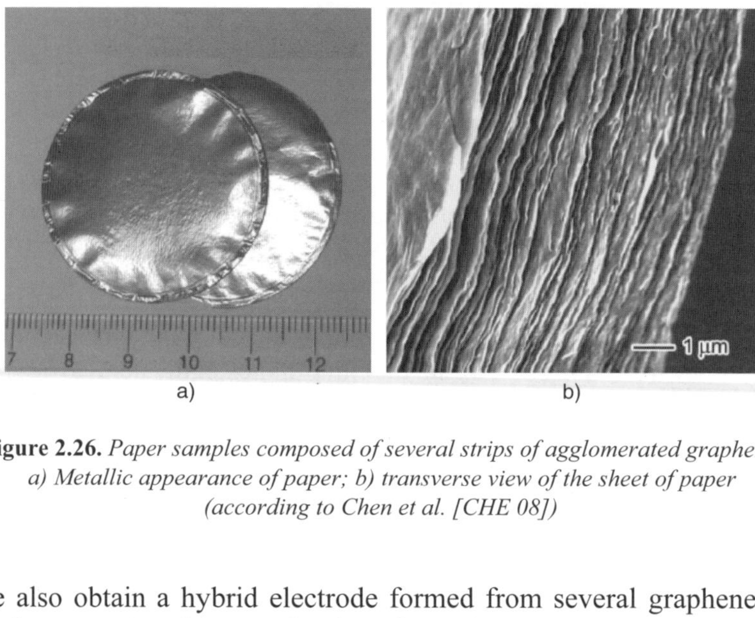

Figure 2.26. *Paper samples composed of several strips of agglomerated graphene. a) Metallic appearance of paper; b) transverse view of the sheet of paper (according to Chen et al. [CHE 08])*

We also obtain a hybrid electrode formed from several graphene layers (equivalent to a metal layer of only a few microns) on which a network of perfectly arranged gold NPs are deposited (Figure 2.27).

As we will see later (in Chapter 3), fine gold particles are much better catalysts than gold in its solid state.

These new properties make this type of electrode particularly useful in diverse applications in electronics or medicine where diverse analysis devices are still in great demand, and also in energy, where problems with storage and conversion of natural electrochemicals are hugely dependent on the reactivity of the electrode surfaces.

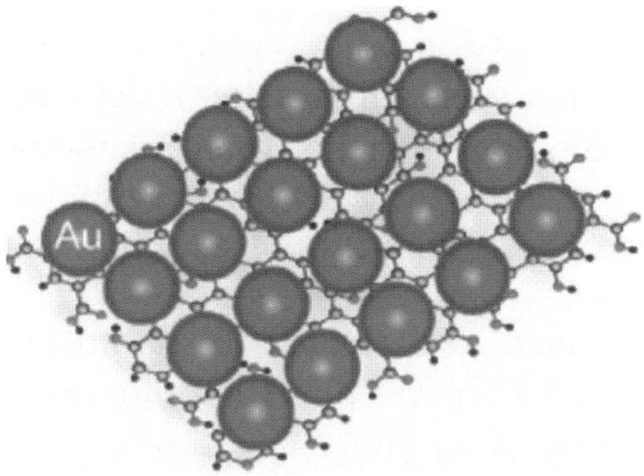

Figure 2.27. *Hybrid electrode of gold graphene nanoparticles. The gold NPs are equally attached to graphene hexagons and provide much improved catalytic properties on the electrode surface compared to a conventional gold electrode (according to Xiao et al. [XIA 12])*

2.2.5. *Link between conjugated carbonaceous nanomaterials*

All these carbonaceous nanomaterials show that they belong to the same family, due to their identical chains of carbon atoms, in the same sp^2 electronic configuration (Figure 2.28).

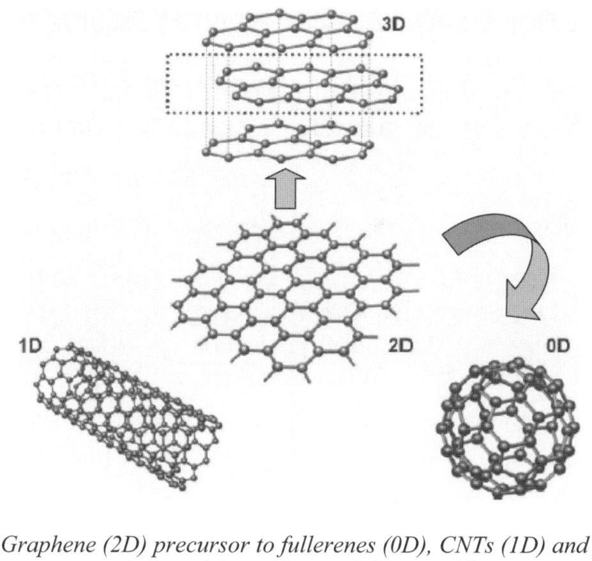

Figure 2.28. *Graphene (2D) precursor to fullerenes (0D), CNTs (1D) and graphite (3D) (adapted from Wan et al. [WAN 12])*

We can also consider that 2D graphene is the first stage and thus a precursor, giving rise to the following: fullerenes (0D), 1D nanotubes and three-dimensional (3D) graphite.

The field of applications of these materials is incredibly fruitful, and there is still a long way to go before all the possibilities are exploited. We can predict rather unexpected discoveries in the near future, with nanotubes and graphene underlying the new properties.

2.3. Conclusions

The study of solid surfaces developed in the 1960s, but no reference had been made to nanosciences, and yet physical phenomena were well analyzed at the nanometric scale. This period, marked by the introduction and rapid expansion of microelectronics saw manufacturing techniques of thin material layers develop, obtained via lithography in a vacuum, with the resulting systematic studies targeting increasingly smaller objects.

The definition of the term nanotechnology was first put forth in the 1970s, and the amount of research dedicated to the field of nanotechnology has continued to rise ever since. This field has benefited, on the one hand from developments in observation techniques such as TEM and SEM, and the ground-breaking invention of near-field microscopes (STM and AFM) and, on the other hand, from the discovery of new families of carbonaceous nanomaterials, the fullerenes, carbon nanotubes and recently graphene, materials with special properties that have provided almost unlimited fields of research with innovative applications in electronics, energy and medicine.

2.4. Bibliography

[BAE 10] BAE S., KIM H., LEE Y., XU X., PARK J.-S., ZHENG Y., BALAKRISHNAN J., LEI T., KIM H.R., SONG Y.I., KIM Y.J., KIM K.S., OZYILMAZ B., AHN J.H., HONG B.H., IIJIMA S., "Roll-to-roll production of 30-inch graphene films for transparent electrodes", *Nature Nanotechnology*, vol. 5, pp. 574–578, 2010.

[BAI 11] BAI H., LI C., SHI G., "Functional composite materials based on chemically converted grapheme", *Advanced Materials*, vol. 23, pp. 1089–1115, 2011.

[BOZ 99] BOZZOLA J.J., RUSSELL L.D., "Transmission Microscope", "Scanning Electron Microscope", Chapters 6 and 7 of *Electron Microscopy: Principles and Techniques for Biologists*, 2nd Edition, Jones & Bartlett, 1999.

[CHE 07] CHEN Z., LIN Y.-M., ROOKS M.J., AVOURIS P. "Graphene nano-ribbons electronics", *Physica E: Low-dimensional Systems and Nanostructures*, vol. 40, pp. 228–232, 2007.

[CHE 08] CHEN H., MÜLLER M.B., GILMORE K.J., WALLACE G.G., LI D., "Mechanically strong, electrically conductive, and biocompatible grapheme paper", *Advanced Materials*, vol. 20, pp. 3557–3561, 2008.

[GAO 09] GAO W., ALEMANY L.B., CI L., AJAYAN P.M., ''New insights into the structure and reduction of graphite oxide'', *Nature Chemistry*, vol. 1, pp. 403–408, 2009.

[GEI 07] GEIM A.K., NOVOSELOV K.S., "The rise of graphene", *Nature Materials*, vol. 6, p. 183, 2007.

[HAM 07] HAMON L., PASTRE D., DUPAIGNE P., LE BRETON C., LE CAM E., PIETREMENT O., "High-resolution AFM imaging of single-stranded DNA-binding (SSB) protein/DNA complexes", *Nucleic Acids Research*, vol. 35, no. 8, p. e58, 2007.

[HLA 00] HLA S.W., BARTELS L., MEYER G., RIEDER K.H., "Inducing all steps of a chemical reaction with the scanning tunneling microscope tip: towards single molecule engineering", *Physical Review Letters*, vol. 85, p. 2777, 2000.

[KIS 08] KISIELOWSKI C., FREITAG B., BISCHOFF M., VAN LIN H., LAZAR S., KNIPPELS G., TIEMEIJER P., VAN DER STAM M., VON HARRACH S., STEKELENBURG M., HAIDER M., MULLER H., HARTEL P., KABIUS B., MILLER D., PETROV I., OLSON E., DONCHEV T., KENIK E.A., LUPINI A., BENTLEY J., PENNYCOOK S., MINOR A.M., SCHMID A.K., DUDEN T., RADMILOVIC V., RAMASSE Q., ERNI R., WATANABE M., STACH E., DENES P., DAHMEN U., "Detection of single atoms and buried defects in three dimensions by aberration-corrected electron microscopy with 0.5 Å information limit", *Microscopy and Microanalysis*, vol. 14, pp. 454–462.

[KRO 85] KROTO H.W., HEATH J.R., O. BRIEN S.C., CURL R.F., SMALLEY R.E., "C60: Buckminsterfullerene", *Nature*, vol. 318, pp. 162–163, 1985.

[KRO 96] KROTO H.W., Symmetry, space, stars and C60, Nobel Lecture, 1996.

[KUG 09] KUGLER V.M., BEAN S.J., "The benefits associated with a 1 mm beam gas path length on the accuracy of X-ray analysis in the variable pressure SEM", *Microscopy and Microanalysis (Zeiss)*, 2009, available at www.zeiss.com/C1256E4600307C70/EmbedTitelIntern/MM2009Panel/$File/MM_2009.pdf.

[LIJ 91] LIJIMA S., "Helical microtubules of graphitic carbon", *Nature*, vol. 354, pp. 56–58, 1991.

[LIN 10] LIN Y.M., DIMITRAKOPOULOS C., JENKINS K.A., FARMER D.B., CHIU H.-Y., GRILL A., AVOURIS PH., "100-GHz transistors from wafer-scale epitaxial graphene", *Science*, vol. 327, p. 662, 2010.

[LOI 01] LOISEAU A., *Les nanotubes. Matériau du futur*, available at www.onera.fr/ conferences/nanotubes/trans/15.htm.

[NOV 04] NOVOSELOV K.S., GEIM A.K., MOROZOV S.V., JIANG, D., ZHANG, Y., DUBONOS, S.V., GRIGORIEVA, I.V., FIRSOV, A.A., "Electric field effect in atomically thin carbon films", *Science*, vol. 306, pp. 666–669, 2004, available at www.sciencemag.org.

[ROH 86] ROHRER H., BINNIG G., Scanning tunneling microscopy. From birth to adolescence, Nobel Lecture, 1986.

[RUS 86] RUSKA E., The development of the electron microscope and of early electron microscopy, Nobel Lecture, 1986.

[WAN 12] WAN X., HUANG Y., CHEN Y., "Focusing on energy and optoelectronic applications: a journey for graphene and graphene oxide at large scale", *Accounts of Chemical Research*, vol. 45, no. 4, pp. 598–607, 2012.

[XIA 12] XIAO F., SONG J., GAO H., ZAN X., XU R., DUAN H., "Coating graphene paper with 2D-assembly of electrocatalytic nanoparticles: a modular approach toward high-performance flexible electrodes", *ACS Nano*, vol. 6, pp. 100–110, 2012.

[ZEG 11] VAN ZEGHBROECK B., "Review of modern physics", Chapter 1 in *Principle of Semiconductor Devices*, 2011, available at http://ecee.colorado. edu/ ~bart/book/book/chapter1/ch1_2.htm.

Chapter 3

Nanomaterials in All Their Forms: New Properties Due to the Confinement of Matter

Alongside the current research into carbonaceous nanomaterials (fullerenes, CNT and graphene), the chemical synthesis of nanomaterials in the form of various types of nanoparticles (NPs), nanowires (NWs), nanotubes (NTs) based on elements other than carbon, or more traditional carbonaceous compounds (polymers), has greatly developed, particularly due to their use in new emerging techniques for diagnosis and medical therapy.

Indeed, chemists and biologists are now able to synthesize multifunctional NPs, to which a variety of molecules bond. These molecules can exhibit therapeutic properties, and the particle–molecule unit is then designed to act specifically on damaged cells; they can also be used in the diagnosis of an illness, i.e. the molecule interacts with a specific region of the body producing a signal that indicates the malady.

These extremely innovative and promising research pathways are currently being put into action for some maladies such as cancer, diabetes and cardiovascular diseases (see Chapter 7).

Semiconductors (SCs) or even polymers can be elaborated with metals; the properties of these nanomaterials are greatly modified when their size decreases and reaches nanometric dimensions[1] (this is the confinement effect of matter). As we will see in Chapter 4, there are two main causes underlying these new properties:

1) a general effect due to the increasing number of atoms on the surface with regard to the bulk atoms;

2) a specific effect, particularly important in the case of metals and SCs, due to modifications in the material's electrical properties. This effect is explained in more detail later on in the case of gold NPs and semiconducting NPs corresponding to Quantum Dots (QDots).

3.1. The different types of nano-objects: main methods of preparation

Currently, there are numerous techniques for elaborating NPs; they can be applied to a large variety of materials, particles existing freely in the form of a powder, others that can be dispersed in the form of suspensions in a liquid (colloids), or even incorporated into a polymer matrix to form reinforced hybrid materials.

They can also be deposited onto a surface to form 2D or 3D NPs arrays (several layers), giving them new physical properties, particularly new optical properties.

3.1.1. *Colloidal solutions of gold NPs*

Chemically manipulating gold to form NPs is without doubt one of the oldest ways to create them. Such preparations seem to have been around since ancient times; glaziers in the Middle Ages used gold NPs to color the glass in cathedrals. However, it was not until 1857 that Faraday discovered that their properties were due to the extreme division of the metal.

Preparing them as a suspension in solution is very simple and can be carried out by anyone in possession of these two base reactants: $NaAuCl_4$

1 For sizes less than or equal to 1 nm, we are no longer talking about an NP but an aggregate of a particular number of atoms (cluster). Au_{55} therefore represents an aggregate of 55 gold atoms with a size of 1.4 nm. These aggregates have very different properties from NPs of a larger size and can be extremely toxic to an organism [SCHMID 08]. They also exhibit fluorescence properties that disappear as soon as NPs increase from 2 to 3 nm.

(sodium tetrachloroaurate), a gold salt, and a reducing salt (sodium citrate), both in an aqueous solution, in an acid medium.

Mixing these two solutions leads to the formation of gold NPs, whose size and function reflect the respective concentrations of the reactants. The gold particles suspended in the liquid are translucent (because of their size being much smaller than the wavelengths of visible light) making up a colloidal solution.

Sodium citrate plays a double role in the formation of gold NPs. On the one hand, it reduces and precipitates the Au^{3+} ions into Au° (metallic gold), and, on the other hand, it adsorbs onto gold atoms located around the edge of the particle (forming a weak energy bond with the gold atoms).

By adsorbing onto the gold atoms, it forms a protective layer with negative electric charges (from the $-COO^-$ groups of citrate ions) that causes repelling forces to be exerted on the other particles, preventing them from forming aggregates.

The solution is initially orange in color, but as the reaction proceeds, it takes on a red color, as a result of the extreme division of gold into NPs of approximately 20 nm in diameter.

The size of NPs can be changed according to different concentrations and temperatures, and hence a wide range of distinct colors from blue to red can be obtained.

The solutions, obtained with sodium citrate, are very sensitive to any variation in acidity and salt concentrations. Therefore, it is preferred to use other molecules than citrate anions as stabilizing agents so that these NPs are more stable.

Sulfur derivatives that bond strongly with gold are most commonly used as they can easily substitute the citrate reagents that are bonded more weakly to gold.

It is also possible to completely remove the citrate ions using sulfur derivatives in conjunction with a reactant based on boron (sodium borohydride, $NaBH_4$); this transforms the gold salts into metallic gold (Au°).

This produces gold NPs surrounded by a layer of organic compounds, bonded to gold via sulfur (S). The NPs are then dissolved in an organic solvent forming a suspension, with the hydrocarbon chains bonded to the sulfur atoms (Figure 3.1).

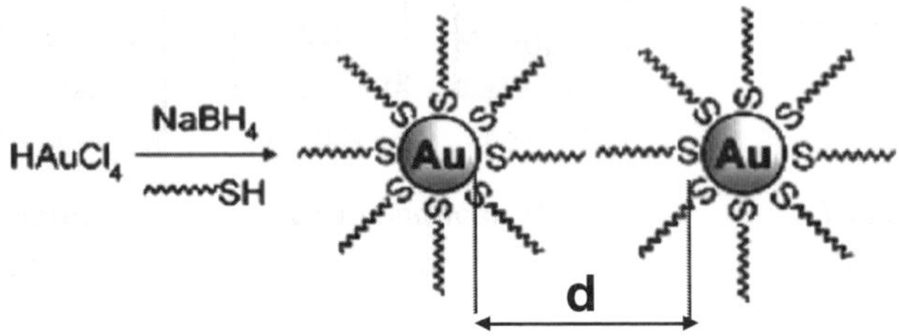

Figure 3.1. *Simplified diagram showing gold NPs obtained by the reduction of an acidic gold solution ($HAuCl_4$) by sodium borohydride ($NaBH_4$) and stabilized by sulfur-based molecules (thiols). The distance* ***d,*** *defined approximately by the chain length of the two sulfur derivatives, corresponds to the minimum distance two NPs can be from one another and corresponds to a kind of physical barrier preventing the NPs from clumping together (adapted from Daniel and Astruc [DAN 04])*

3.1.2. *Hybrid and magnetic NPs (ferromagnetic fluids)*

Using iron-based magnetic particles to make ferrofluids is not a new idea. It was patented in the 1960s in the USA and it was developed in the 1980s in France by Massart [MAS 81].

Preparing a ferrofluid is relatively simple. By mixing two ferrous and ferric salt solutions or by adding sodium hydroxide or ammonia to the mixture, the ferrofluid is produced in the form of Fe_3O_4, a suspension of oxide particles in the solution.

The solution remains stable if the particle size is smaller than 10 nm. At room temperature, thermal agitation, sufficient to resist the magnetic forces of attraction, prevents the NPs from coagulating and clumping together. By using suitable surfactants, one can also obtain NPs with sizes greater than 10 nm without any aggregates forming.

The physicochemical properties of these ferrofluids are amazing; they have been the subjects of numerous applications across many different fields (optics, mechanics and more recently in the biomedical field).

The possibility of confining these iron oxide NPs (Fe_2O_3 or Fe_3O_4) ranging between 8 and 100 nm in specific regions by using a magnetic field is rather interesting. It has already been the subject of important applications in medicine for over a decade. This property of confinement is in fact providing a solution to the problem of targeted vectorization of drugs, as well as to the rapid removal of reagents.

However, it appeared that the chemical instability of iron oxides, particularly in weaker acidic solutions, was a problem. Also, in order to overcome this disadvantage, the iron oxide NPs can be protected by a thin layer of an inert substance such as silica or gold, leading to "core/shell"[2] particles where iron oxide occupies the core of the particle.

A protective shell of gold has numerous advantages. The NP not only conserves its magnetic properties, but also acquires the properties of gold NPs; this considerably increases the possible applications in the field of medicine.

Preparing these core/shell hybrid NPs involves two stages: from colloidal solutions of iron oxide NPs obtained from sodium hydroxide and a mixture of iron chlorides and then reaction in the presence of gold salts and a catalyst allowing the iron oxide NPs to be coated in a thin film of gold, whose thickness can easily be controlled.

A similar technique is used to make a silica shell, which also has numerous advantages, in particular its lack of toxicity. Moreover, these hybrid NPs can also be modified by bonding to a variety of molecules (Figure 3.2).

The controlled deformation of the ferrofluid surface under the effect of a magnetic field is also a specific property, which can give rise to rather spectacular effects, such as the formation of peaks on the surface of the ferrofluid giving it the appearance of a sea urchin (Figure 3.3).

This last effect is visible only under rather specific conditions: a particular viscosity of the solution, the size of the NPs, their concentration and the strength of the magnetic field.

---

2 The symbol core/shell is often used to describe a hybrid NP. Thus, Fe_2O_3/Au means that Fe_2O_3 is the core of the particle and Au is the shell.

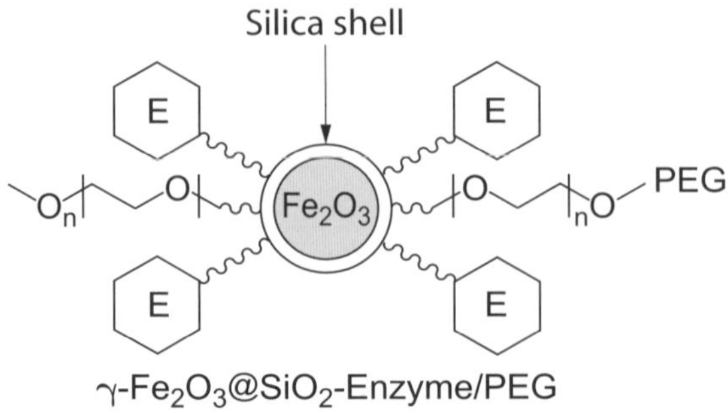

Figure 3.2. *Example of a ferromagnetic core/shell NP with a Fe_2O_3 core protected by a silica shell (SiO_2) to which are bonded, on the one hand, an enzyme (E) with therapeutic properties and, on the other hand, a long hydrophilic chain of polyethylene glycol (PEG), whose function is to make the NPs soluble in the aqueous medium (adapted from Georgelin et al. [GEO 10])*

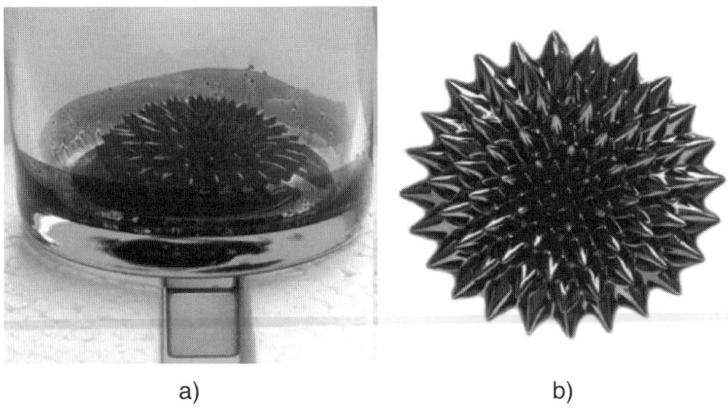

Figure 3.3. *Deformation of the surface of a ferrofluid (nanoparticles of 10 nm) under the effect of a magnet beneath the container: a) Side view; b) top view (Images courtesy of Supermagnete: www.supermagnete.fr/FER-01)*

3.1.3. *Semiconducting NPs (quantum dots)*

The electronic structure of SCs in the solid state, different from that of metals, gives them particular opto-electronic properties, at the origin of numerous applications (transistors, electroluminescent diodes, lasers, etc.).

When the size of SCs is within the range of several nanometers, the confinement of matter gives rise to a new energy distribution of electrons providing a large modification in the opto-electronic properties compared to those of the SC in the solid state.

These NPs, identified as QDs or QDots, also exhibit fluorescent properties (see Box 4.2, luminescence) that depend on their size, as observed, for example, with SCs such as cadmium sulfide (CdS) and selenium sulfide (SeS), giving off different colors.

Moreover, the fact that these NPs can be chemically "decorated" with various reactants, capable of interacting with living cells, makes them very valuable diagnostic tools; their use is becoming increasingly more widespread throughout the field of experimental medicine [OGL 11].

Unlike certain organic fluorescent molecules that are also used as biomarkers in diagnostic tests, these NPs have numerous advantages. They are in fact highly chemically stable; the intensity of the fluorescence weakens only slightly over time, which allows relatively long observation times *in vivo*. Finally, a full range of fluorescent colors can be obtained from the same compound by simply modifying the size of the NPs (see Chapter 4).

The fluorescence is also detectable at extremely low intensities. Using QDots as vectors for transporting diagnostic reagents is becoming more and more widespread and tends to replace organic fluorescent dyes.

From a practical point of view, we must note however that the SC used, CdS or CdSe, are highly toxic compounds, and this effect must be minimized before they are injected into the organism.

A solution involves coating the NPs in another SC such as zinc sulfide, less toxic than CdS or SeS. The colors emitted by CdS or CdSe are preserved, and the thin layer of ZnS deposited on the particle core increases the intensity of the fluorescence even more.

By absorbing photons of energy $h\nu$ (excitation step), an electron is excited from the energy level E_0 (fundamental state) toward a higher energy level E_1 (excited state). This excitation step is extremely quick (around 1 fs, i.e. one a billionth of a millionth of a second or 10^{-15} s).

Box 3.1. Luminescence

Certain substances, after absorbing energy, can luminesce, i.e. emit light. Among the different types of luminescence (chemo-, bio- and electro-luminescence), fluorescence and phosphorescence are the two kinds of light emitted and observed after matter absorbs light radiation.

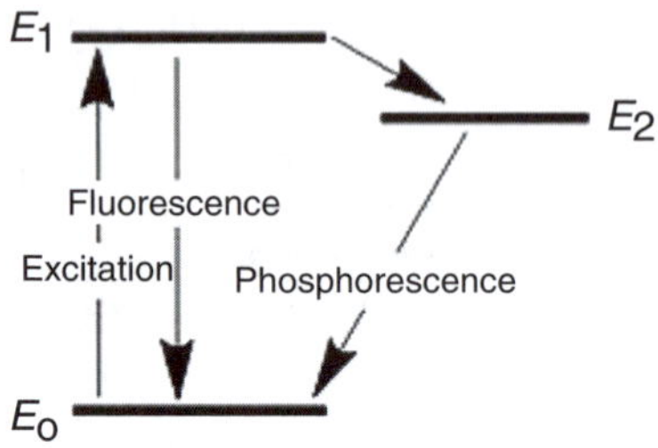

The first possibility is a deactivation in the excited state of the electron that returns to E_0 via collisions with neighboring atoms. No photons are emitted; it is a simple transformation of energy into heat.

The second possibility corresponds to an emission of photons and to the fact that the excited state E_1 is much greater in length, lasting approximately 10 billionths of a second (10^{-8} s). This is the phenomenon of *fluorescence*, characterized in the case of components in solution by light emitted with a wavelength λ_F, that is slightly higher than the wavelength of excitation λ_o. The fluorescence disappears as soon as excitation ceases.

The third possibility is much rarer, and corresponds to an energy transition from E_1 toward an intermediate level E_2. It is much slower, and causes a change in the configuration of the spin of the electron (see Chapter 5). The transition from E_2 back to E_o involves the emission of light, with a wavelength λ_P greater than λ_F: this corresponds to *phosphorescence*. The life time of state E_2 is longer (from 1 ms to several seconds), and this explains why phosphorescence is still present for some time after the excitation has stopped.

Just like for the gold NPs, these core/shell QDots can be modified by adding numerous reactants onto its surface; this gives them multiple functions (Figure 3.4).

First of all, bonding hydrophilic molecules or polymers onto the surface makes them compatible with the surrounding biological environment. NPs also enable us to take images *in vivo* of body regions affected by an illness; the QDots are "decorated" by reagents capable of selectively adhering onto damaged cells or cells affected by the illness. This allows us to identify the affected regions with precision, by directly observing the fluorescence of the NPs in situ.

The whole point of this technique is to make possible the early detection of an illness so that an effective therapeutic treatment can be adapted to the patient and hence the chances of recovery of the patient can be increased (see Chapter 7).

Figures 3.4 and 3.5 show some examples of molecules or polymers capable of bonding to QDots.

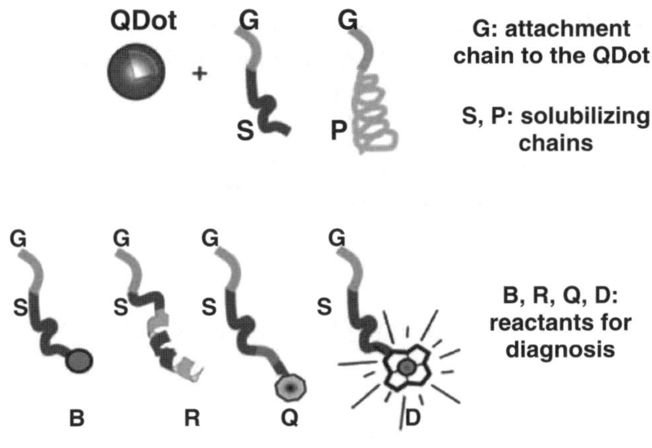

Figure 3.4. *Some examples of molecules that can bond to a QDot. Each molecule is formed from the same starting block G, responsible for attaching the QDot to the surface. The molecules S or P assure the solubility, the specific groups B, R, Q and D linked to S provide the diagnostic functions. B, biotin, is a protein that can bond very tightly to another protein such as streptavidin; R is a specific peptide chain capable of recognizing certain proteins. Q is a "quencher" or a group of molecules that prevents the QDot to fluoresce; this is regenerated after the peptide chain linking Q to S is broken. D is a chelator of radioactive elements (a molecule that can immobilize radio elements). Joining a radio element and a QDot together allows a double imagery to be produced, due to the fluorescence of the QDots and the radioactivity given off by the radio element (adapted from Michalet et al. [MIC 05]). For a color version of this figure see www.iste.co.uk/lacaze/nano.zip*

Notice that these molecules all have a component in common, G, which allows them to bond to the QDot (usually via a –SH bond). Chains of S or P make them soluble in the biological environment, whereas, a specific group of B, R, Q or D gives them specific diagnostic properties.

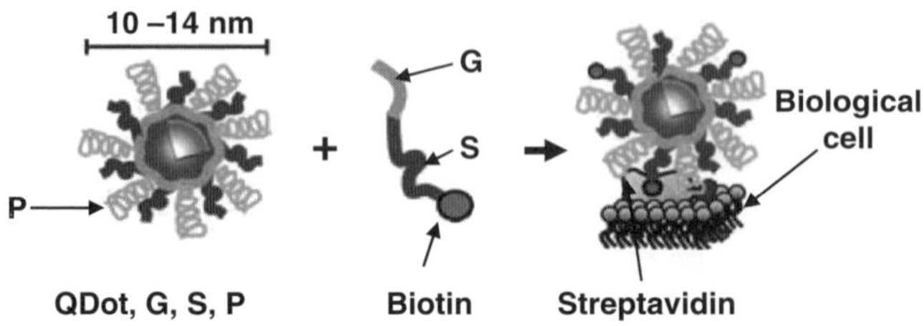

Figure 3.5. *Attachment of a QDot onto a biological cell. The QDot is chemically modified by biotin and rendered hydrophilic once bonded to molecules G–S and G–P. The biotin molecules attached to the surface of the QDot react selectively with streptavidin molecules prefixed onto biological cells (adapted from Michalet et al. [MIC 05]). For a color version of this figure see www.iste.co.uk/lacaze/nano.zip*

3.1.4. *Phospholipid vesicles and encapsulation by liposomes*

These are NPs whose size can vary between 10 nm and several hundred nanometers, spontaneously forming in water via a self-assembly process between amphiphilic molecules, similar to surfactants. These molecules have a hydrophilic head linked to a hydrophobic hydrocarbon chain, soluble and insoluble in water, respectively.

The only way these molecules can exist in the solution is to group them together and form a supramolecular structure as stable as possible, corresponding to a minimal energy. This is possible when the hydrophilic heads are in contact with water molecules and the hydrophobic chains group together to escape the forces of repulsion exerted by the water molecules. Therefore, supramolecular structures are formed with weak cohesion, making them easily deformable.

Depending on the concentrations of molecules in solution, extremely varied assemblies can be obtained; for example, lamellar, tubular or even spherical vesicles or liposomes.

Biological membranes of living cells are composed of a phospholipid bilayer. This maintains a watertight seal between two distinct aqueous mediums, while allowing ionic exchanges between the interior and the exterior of the cell via specific proteins inserted into the membrane.

The pharmaceutical industry uses liposomes and vesicles fairly regularly to form the capsules of pills, which are often hydrophobic; on the one hand, this allows them to be dissolved in aqueous biological mediums, and, on the other hand, they are prevented from breaking down too early when they enter an organism.

The use of non-toxic natural amphiphiles that can form various sizes of liposomes is particularly advantageous. Lecithin phospholipids such as phosphatidylcholine are the most commonly used phospholipids.

Characterized by a double hydrophobic chain of hydrocarbons derived from natural fatty acids (palmitic and oleic acid), it is joined to a hydrophilic head with a negatively charged phosphate group and a choline group with a positively charged nitrogen atom.

These electrical charges carried by the phosphate and nitrogen groups are largely responsible for the hydrophilic properties of the molecule (Figure 3.6).

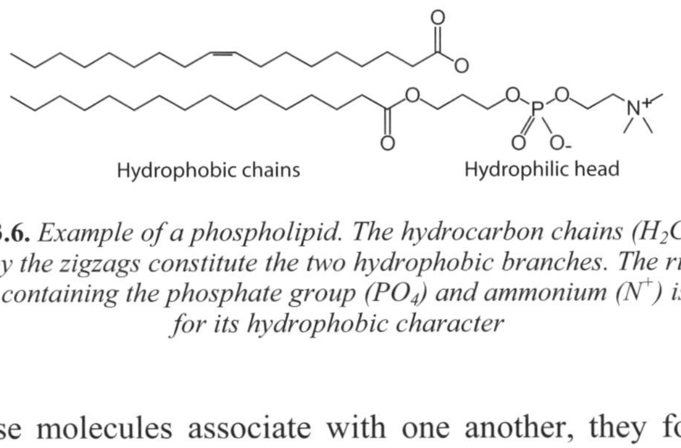

Figure 3.6. *Example of a phospholipid. The hydrocarbon chains (H_2C–CH_2–) indicated by the zigzags constitute the two hydrophobic branches. The right part of the molecule containing the phosphate group (PO_4) and ammonium (N^+) is responsible for its hydrophobic character*

When these molecules associate with one another, they form spherical liposomes under certain concentrations, with two concentric layers of phospholipids linked to one another by their hydrophobic parts; the internal and external surfaces of the sphere are composed of the hydrophilic heads (Figure 3.7).

These liposomes are able to deform while maintaining cohesion; this allows them to encapsulate hydrophobic (or lipophilic) components that will be inserted into the area defined by the hydrocarbon chains. This property has been used for a long time in pharmacy to transport certain drugs into the organism and to target damaged or cancerous cells more effectively.

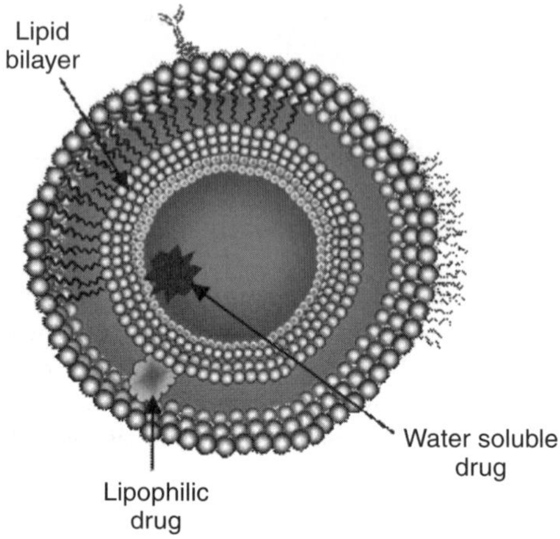

Figure 3.7. *Diagrammatic representation of a liposome (cut in half) formed from an assembly of phospholipids. The hydrophilic heads represented by little balls are in contact with water, whereas the hydrophobic chains are clustered together in such a way that they form a hydrophobic bilayer isolated from the water*

For example, vitamin C encapsulated by liposomes based on phospholipids is currently available on the market. It is much more efficient compared to vitamin C ingested as tablets with mineral coatings. It can in fact circulate freely around the organism without being prematurely broken down.

Encapsulation by phospholipids has other advantages too.

In the case of diseased cells whose walls are also composed of phospholipids, they can have the tendency to capture the phospholipid molecules in the liposome, thus releasing the drug in the place where it is required to act. This mechanism, underlying numerous therapies, allows the same therapeutic effect to be obtained by accurately calculating doses of medicines while avoiding undesirable secondary effects.

3.1.5. *Nanowires*

The interest in these types of materials is fairly recent, and since the year 2000 they have been studied in much more detail by several teams.

Techniques to develop them have multiplied and it is now clear how to create NWs of only a few nanometers in diameter and several hundred microns long, regularly arranged into well-adapted networks for making electronic "chips" with diverse functions.

The growing interest in these NWs is nicely shown by the dramatic increase in the number of scientific articles dedicated to this subject that were published worldwide between 1991 and 2009 (Figure 3.8).

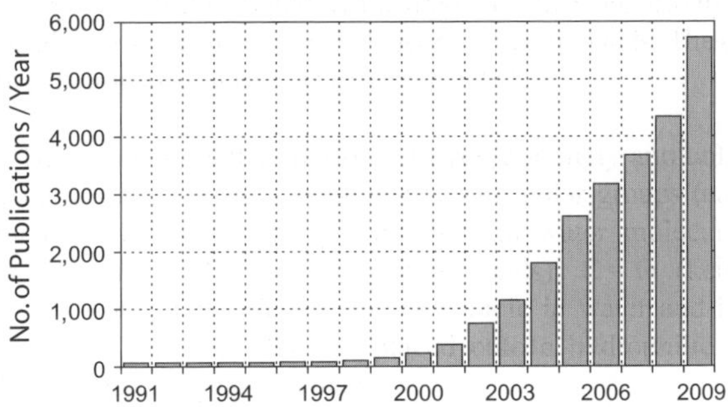

Figure 3.8. *Evolution in the number of publications dedicated to the study and to the applications of nanowires between 1991 and 2009 (adapted from Yang et al. [YAN 10])*

Studying NWs could have been considered as a laboratory curiosity (only about a hundred publications dealing with NWs in 1998), but it has rapidly become a major field of research (over 5,500 publications in 2009). This "exponential" increase is not a fad, but instead can be explained simply by the sheer number of applications resulting from it.

Currently, out of all the studies on NWs, those based on silicon are the ones still interesting researchers. Their electronic properties are very different from those of solid silicon and have led to the creation of field-effect nanotransistors. These act as ultrasensitive sensors; this allows them to be used in applications in diverse areas of research, particularly in the detection of traces of chemical components (see Chapter 7).

The production of isolated NWs is not very useful. Instead, it is preferred to duplicate them by depositing them onto an insulating base, in parallel with one another; they are separated by only a few microns, which allows nanoscale "chips" in transistors to be produced.

Among the different fabrication techniques, one of them, known as superlattice nanowire pattern transfer (SNAP), is very powerful; it was developed in 2003 by several Californian research groups (Melosh *et al.* [MEL 03]).

It is a manufacturing process directly inspired by lithography techniques used in microelectronics.

The process involves combining evaporation and arrangement of diverse materials into thin layers under vacuum conditions, followed by controlled chemical pickling. We therefore obtain NWs of between 10 and 20 nm in diameter, arranged onto a solid insulating surface in a regular fashion (usually silicon covered in silica).

They can reach lengths of several tens of microns, which is more than enough to create electronic chips. This technique is quite common and can be applied to a variety of metals (Figure 3.9).

More recently, it has been adapted to plastic substrates, allowing a much greater field of applications to be considered, such as "flexible" or "plastic" microelectronics.

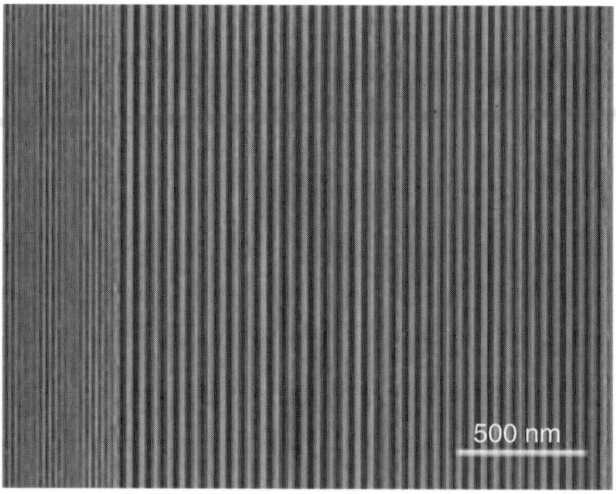

Figure 3.9. *SEM image of Si nanowires (NWs) arranged on a surface of SiO_2. The left part of the figure represents 19 NWs with a length of 18 nm, each one separated from another by spaces of 30 nm; the right part of the figure corresponds to 40 large NWs of 20 nm separated by 60 nm. The continuity and regularity of the NWs are maintained up to 100 µ (adapted from Melosh et al. [MEL 03]*

Progress made in the manufacture of nanometric wires has been rapid and is becoming similar to semi-industrialized techniques.

Even more recently, in 2011, a group of researchers from Bilkent University in Turkey succeeded in creating highly regular NW cables of several dozen nanometers in diameter, but with a length of several tens of meters (Figure 3.10).

Figure 3.10. *Network of nanowires in a semiconductor (As_2Se_3) covered in a thin capsule made up of an insoluble polymer (PVDF or polyvinylidene difluoride). These nanowires, whose diameter can be adjusted from several hundred nm to approximately 10 nm, can be produced with lengths of several dozen meters. They are regularly arranged inside a cylindrical polymer matrix (with a diameter of approximately 60 µ) that can be dissolved in a solvent, thereby recovering single nanowires; their initial parallel orientation is conserved. The images show that the arrangement of nanowires inside the polymer matrix (white dots on the photos) is perfectly conserved at both ends of a fiber with a length of 10 m (adapted from Yaman et al. [YAM 11])*

3.2. Organizing nanoparticles into arrays

Numerous methods have been developed to produce organized arrays of NPs arranged on substrates, by using self-assembly methods or lithographic techniques, completely different and not as expensive as those employed in the microelectronics industry.

Building NP networks, arranged in an orderly fashion on a surface, is particularly interesting due to the appearance of new properties, because of combined effects that can greatly improve or modify the specific properties of each NP.

The most promising techniques are undoubtedly those that take into account the self-assembly properties between different molecular structures. The techniques, usually inspired by the living world, are based on the formation of complex molecular structures corresponding to supramolecules.

The production reproducibility and the order and spacing between particles are key factors determining the future use of their properties.

3.2.1. *Self-assembly*

Self-assembly forms the basis of all processes used to build molecular structures and originates from the "bottom-up" techniques[3].

The idea is to build onto a surface, a thin membranous layer with a regular network of pores in which groups of molecules or NPs in solution may be spontaneously inserted or attached onto the surface.

This is the "template" effect. It is similar to a mask pierced by many tiny holes, which allows molecules or NPs in solution to squeeze through the crevices and to react with the bare areas of the surface. This "template" effect can be achieved in many different ways, and it depends on the desired gap size between particles and, therefore, the number of pores per square centimeter of the membrane that will have been chosen to form the mask.

For extremely small dimensions, ranging from one to several dozen nanometers, masks are obtained via a preliminary assembly of "molecular bricks", i.e. particular polymers or molecules able to adsorb spontaneously onto the surface.

---

3 Surface nanostructuring can be carried out using two techniques: "top-down" and "bottom-up''. Developing transistors in thin layers, which requires a series of operations (pickling, ion implantation and successive depositions of ultrathin layers of materials), is typically a "top-down" technique, i.e. layers formed from top to bottom. The opposite case, where nanostructuring is achieved by directly attaching molecules or nanoobjects onto the bare surface, are "bottom-up" techniques, i.e. the network is constructed from bottom to top.

Moreover, these "molecular bricks" are chosen to self-assemble and form regular networks of grooves on the surface in which NPs will be inserted.

To obtain networks with less compact mesh, aluminum membranes are used (distances between the pores can be adjusted from several nanometers to several hundred nanometers). This type of membrane is very widespread; it is obtained by the electrochemical oxidation of a fine aluminum layer.

Beyond these dimensions, a very different technique is used; it involves arranging a perfectly calibrated compact layer of nanospheres of silica or polystyrene onto a surface. A partial masking of the surface is obtained with regularly distributed gaps between the spheres and in which diverse nano-objects can be deposited.

By choosing the size of the nanospheres, networks, in which the distances between particles will be between hundreds of nanometers and several microns, can be constructed on large surfaces. Even though we leave the "nano" scale here, these networks are very easy to form and possess optical properties similar to those of photonic crystals and, therefore, can be successfully used to make diverse optical devices.

3.2.1.1. *Molecular self-assembly and the formation of nanometric networks*

This is a very promising area of research, and perfectly illustrates the concept of building nanostructured surfaces using the "bottom-up" method. These techniques are interesting because they allow structuration dimensions at the molecular level to be attained and thus are far beyond traditional vacuum lithographic techniques used in the microelectronics industry.

Two main self-assembly pathways are currently being studied: assembly by molecules with distinct functions that spontaneously join together on a surface and form perfectly regular networks of cells, or, assembly by block copolymers, the most widespread technique by far, with many more possibilities which has already given rise to many products.

3.2.1.1.1. Self-assembly of molecules on surfaces

Self-assembly is based on a choice of two specific molecules, both with a great association affinity. This association affinity is a result of the presence of antagonistic chemical groups on each molecule, which spontaneously associate, creating weak energy bonds, which ultimately lead to an extremely regular network of small bare metallic islands on the surface.

The example in Figure 3.11 nicely shows this concept and also how a molecular network with nanometric holes can be obtained by the self-assembly of two particular molecules A and B, with "self-associating" chemical groups such as NH and C = O, N and NH. These bonds (known as "hydrogen bonds") are widespread in chemistry and biology, and are in fact weak energy interactions, which can be likened to electrostatic forces between atoms (Box 3.2).

To simplify this, it can be considered as an attraction between positive and negative electric charges; hydrogen is bonded to N and carries a positive charge, whereas nitrogen (N) and oxygen (O) atoms are negatively charged.

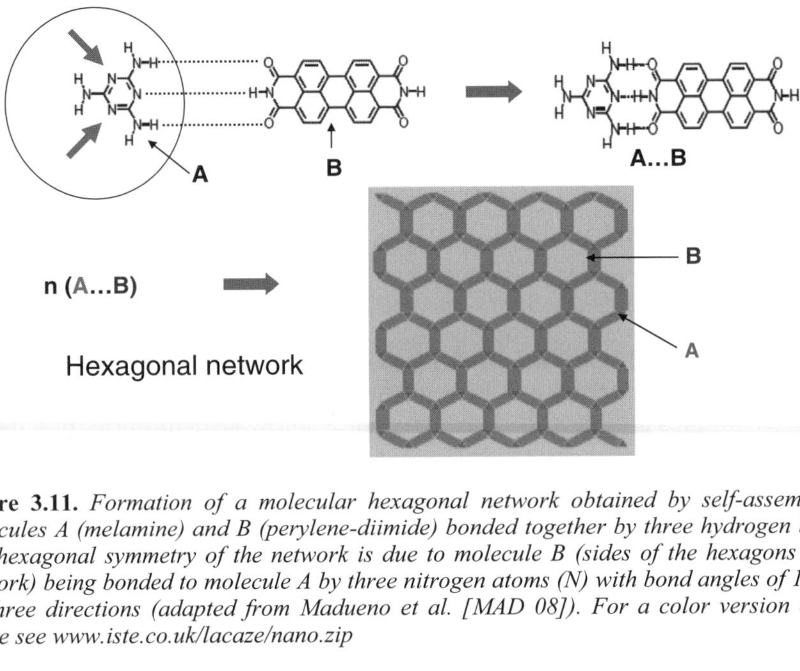

Figure 3.11. *Formation of a molecular hexagonal network obtained by self-assembly of molecules A (melamine) and B (perylene-diimide) bonded together by three hydrogen bonds. The hexagonal symmetry of the network is due to molecule B (sides of the hexagons of the network) being bonded to molecule A by three nitrogen atoms (N) with bond angles of 120° in all three directions (adapted from Madueno et al. [MAD 08]). For a color version of this figure see www.iste.co.uk/lacaze/nano.zip*

In general, such networks are obtained by evaporation of reactants under an ultra-high vacuum; their stability then becomes a problem when they come in contact with the surrounding air.

Box 3.2. Hydrogen bonds

These are weak energy bonds resulting from electrostatic forces of attraction between opposite electrical charges (δ^+ and δ^-).

This situation is frequently found in biology and chemistry and occurs each time a compound with a hydrogen atom with acidic properties (δ^+ charged), such as A–N**H**$^{\delta+}$ or A-O**H**$^{\delta+}$ which is found in the presence of a hydrogen acceptor $^{\delta-}$**B**-Y (δ^- charged), corresponding to groups such as **O** = C–Y, **O** = P–Y and **N** = Y.

The interaction between opposite charges spontaneously leads to a pseudo-bond between H and O, N, called a hydrogen bond (H bond):

A–O**H … O** = C–Y, A–N**H … O** = P-Y, A–N**H … N** = Y

and to the formation of supramolecular structures whose stability increases as the number of hydrogen bonds increases. It was J.M. Lehn (Nobel Prize winner, 1987) who developed this field of chemistry, better known as supramolecular chemistry.

The unusual properties of water (higher boiling and freezing points) are due to the presence of H bonds, as well as the hydrophilic properties (affinity for water) of certain compounds which are also due to the formation of H bonds with water; this makes them soluble in this medium.

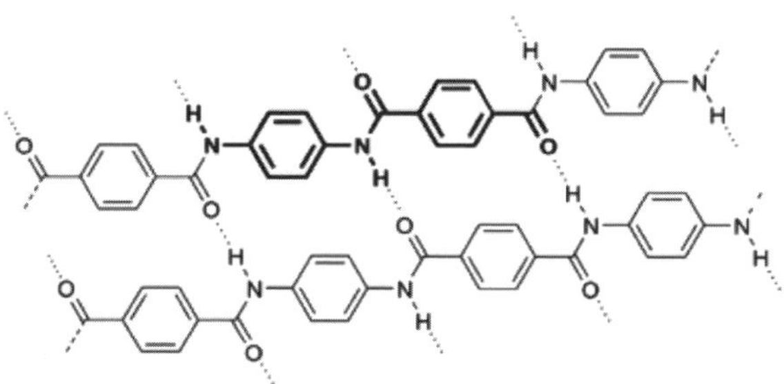

A material such as Kevlar (polyamide) owes its exceptional mechanical properties to the presence of an infinite number of H bonds between C = O and HN groups; this causes the rigidity of the material to increase considerably.

This type of interaction is rather common throughout the living world and plays an early role in the replication phenomena of DNA strands, corresponding to hybridization reactions.

In the present case, the remarkable fact is that this molecular network is obtained in the surrounding air by mixing solutions of different products. The procedure involves dissolving the two compounds A and B in a particular solvent, applying the solution onto a surface, and then leaving the solvent to evaporate. These holes defined by the network are possible sites that NPs or particular reagents can enter.

Undoubtedly, this broadens the horizons for the large-scale construction of cheap nanometric networks. However, we must recognize that we are still a long way from being able to use this procedure in automated production lines. Important research efforts are still needed to increase the reliability of this process. The discovery of new molecular systems, capable of self-assembling onto surfaces through the formation of flawless porous nanometric networks, is currently one of the main objectives of the research.

3.2.1.1.2. Self-assembly of block copolymers

This process is similar to the previous process, and more specifically to a particular family of polymers: block copolymers. Polymers are long-chained molecules achieved by the repetition of the same monomer A, leading to huge molecules such as ...–AAAAAAAA–.... Copolymers have a more complex structure; they are composed of either a regular or a random chain of several monomers. The simplest example of a copolymer AB is produced by linking together A and B. To distinguish copolymers whose monomers A and B are randomly linked, we can write B–AAA–BB–AAAAA–BBBB–A, or contrarily those linked in a regular fashion with homogenous groups, such as ...–BBB–AAAAAA–BBB–AAAAAA–BBB–... called block copolymers, which can be written as $(A_pB_q)_n$, or $(A_6B_3)_n$ if using the previous example.

It is with this last type of block copolymers that interesting self-association properties are observed at the nanometric scale, giving rise to the formation of microdomains with sizes between 5 and 50 nm.

A large number of accomplishments have already been carried out at the nanometric scale in the laboratory and under industrial conditions. Thus, this nanostructuration pathway could be a future solution to expand the current limits of photolithography (currently around 22 nm).

These microdomains are created by following simple rules, essentially based on thermodynamics; this allows their morphology to be predicted.

Microdomains can also be obtained with spherical, cylindrical or lamellar structures or with other more complicated morphologies, such as a so-called gyratory structure (Figure 3.12).

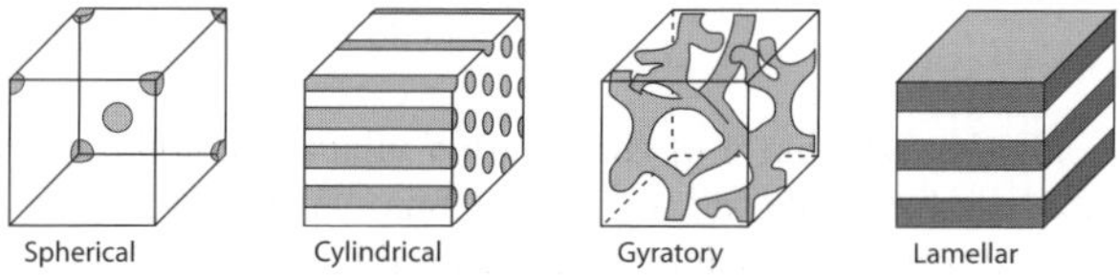

Figure 3.12. *Some examples of different structures of microdomains obtained for a solid block copolymer AB; the proportion of one of the monomers A increases from left to right (A in black, B in gray) (adapted from Bates and Fredrickson [BAT 99])*

These surprising nanostructures come from the multiple folding of chains, brought about by the self-assembly of monomers. This specific self-assembly can be explained by considering that the forces of interaction between identical monomers are preferred to those between different monomers.

Figure 3.13 shows diagrammatically how a lamellar structure can be obtained from a three-block copolymer P–(I–S)$_m$–I–P: P is a long-chain block (poly-2vinylpyridine), and I (polyisoprene) and S (polystyrene) are short-chain blocks; the chain lengths will determine the spacing between the different layers.

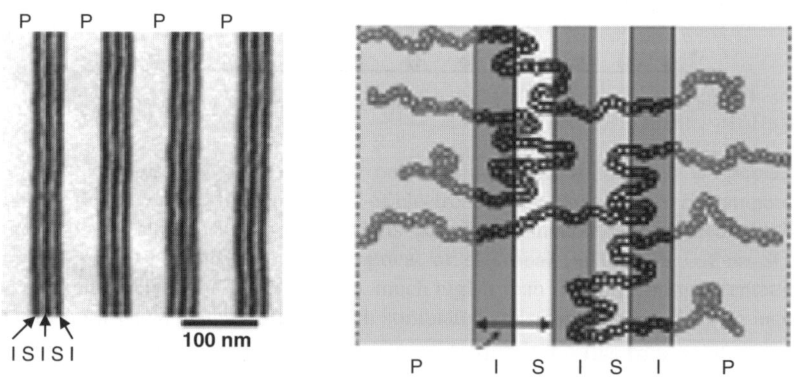

Figure 3.13. *TEM image of a section of a film of the three-block copolymer P–(I–S)$_2$–I–P and a self-assembly diagram of identical blocks I, S and P leading to a regular lamellar structure (photo to the left) (adapted from Bates and Fredrickson [BAT 99])*

Numerous examples of how nanomaterials are structured have been described over the past decade. By synthesizing a large variety of block (2, 3 and more) copolymers, polymer scientists have shown that it is possible to use this technique to develop a multitude of nanomaterials, particularly membrane pores of only a few nanometers in size (Kim *et al.* [KIM 10]). Besides this, another challenge is to be able to accurately position these nanostructures onto a surface and to increase the regularity of the motifs for nanostructuring.

This was recently revealed. Rather astonishingly, by making a preliminary structure of the surface (using a photolithography technique, dimensions of beyond 100 nm), it becomes possible to deposit the copolymer with nanopores in a particular position, and to obtain an extremely regular nanostructuring on areas of only a few microns squared with sizes of approximately 20 nm.

Significant improvements have recently been made in order to produce several electronic components (the components of microprocessors) on a laboratory scale [KIM 10]. It is believed that this area of research is most promising and can be imposed in the future to extend the miniaturization of electronic components beyond 20 nm, particularly because the manufacture costs to produce these electronic components are much less than those involved in traditional photolithography.

3.2.1.2. *Self-assembly of NPs on solid surfaces*

The principle is the same as the previous and is also based on a self-association phenomenon between NPs and molecules, used as mediators during the assembly.

We must remember that these NPs (metallic or semiconducting) are always coated by molecules, which allows them to remain in the solution in the form of a suspension.

These NPs can self-assemble by placing a drop of the solution onto a surface in the presence of a mediator, and by leaving the solvent to evaporate. This mediator is a bifunctional compound, with a reactive group at each end (SH groups in Figure 3.14), capable of displacing a coating molecule and of fixing itself in the same place on the metallic surface. The forces of interaction between the mediator and the NPs are very weak when the NPs are in the solvent. They become much stronger when the solvent is removed, allowing the NPs to link together, according to the diagram in Figure 3.14.

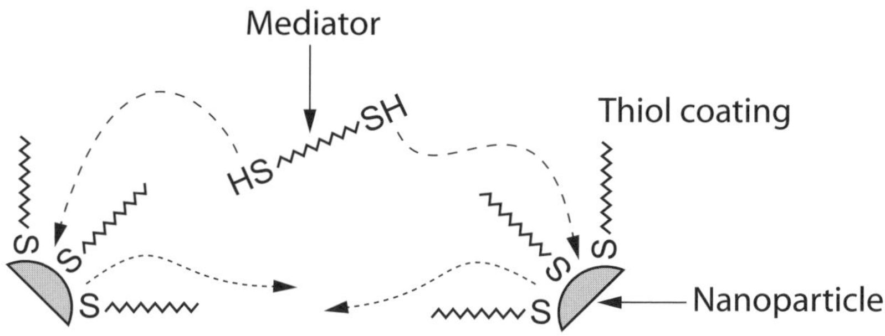

Figure 3.14. *Association mechanism of two gold NPs, carried out via a dithiol compound (HS<><><>SH) used as a bonding mediator. The surfaces of both NPs (shown as semicircles) are coated by thiols (S<><><><>) (adapted from Wang et al. [WAN 10]). For a color version of this figure see www.iste.co.uk/lacaze/nano.zip*

It is thus possible to move from a disordered system (a colloidal solution of NPs) toward a self-organized surface network, where the NPs are regularly arranged according to one another once the solvent has evaporated (Figure 3.15).

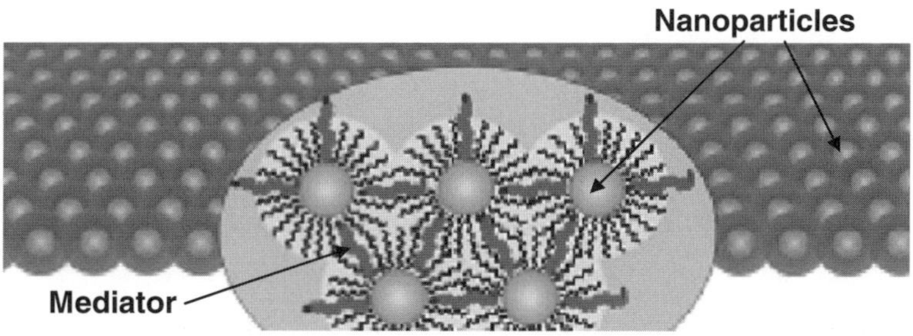

Figure 3.15. *Diagrammatic representation of a self-assembled network of NPs on a surface (adapted from Wang et al. [WAN 10]). For a color version of this figure see www.iste.co.uk/lacaze/nano.zip*

The previous example is a typical assembly diagram. There are other possibilities that have been tested by using diverse molecules as mediators: polymers, proteins, viruses and oligonucleotides (short chains of DNA, Box 3.3), as shown in Figure 3.16.

Box 3.3. DNA

DNA or deoxyribonucleic acid is a macromolecule, the building block of life and inheritance. It is found in a condensed form within chromosomes, inside the nuclei of all cells. It is formed by the supercoiling of two complementary strands of DNA, forming what looks like a twisted ladder; each bar is composed of two purine bases linked together via hydrogen bonds, the only couples possible being adenine–thymine (A–T) and cytosine–guanine (C–G). The 3D structure of DNA was discovered by Watson and Crick, who were awarded the Nobel Prize in 1960.

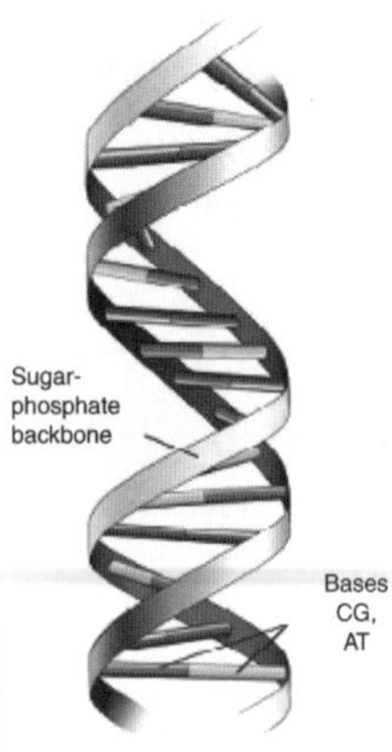

DNA contains the genetic code, determined by the successive order of base pairs in the DNA strand. The code is read in "words" corresponding to a set of three bases, called "codons". Each codon codes for a specific amino acid, then ultimately a protein.

When a strand of DNA carries a sequence of purine bases, i.e. CTGAGTCA, it spontaneously associates with a complementary strand GACTCAGT (hybridization), thus forming a double-stranded fragment of DNA:

```
C T G A G T C A
| | | | | | | |
G A C T C A G T
```

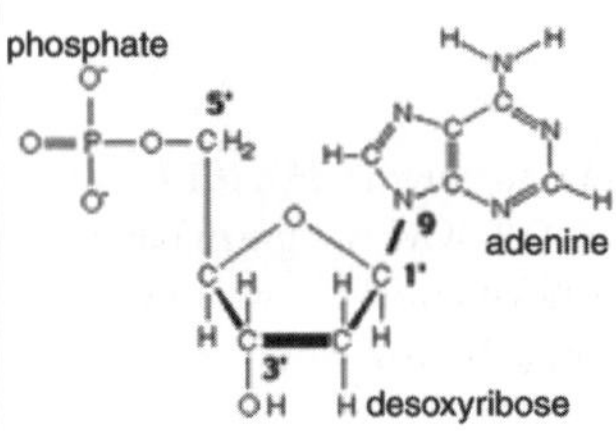

A strand of DNA is formed by a chain of nucleotides; each nucleotide is made up of a phosphate group, a sugar (desoxyribose) and a base (A, T, C or G).

Using grids of DNA as a base has also been proposed to form 2D networks of gold NPs. The carefully chosen strands of DNA are in fact capable of self-associating by producing regular geometric 2D or 3D structures; these structures have been compared to origami (Winfree *et al.* [WIN 98]).

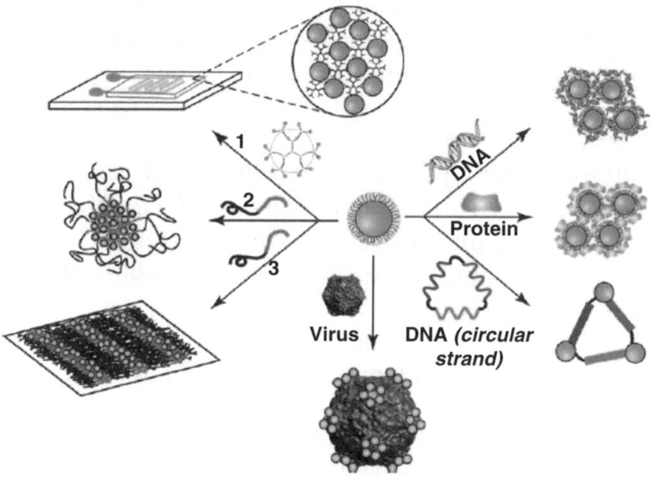

Figure 3.16. *Forming networks of gold NPs by using segments of DNA, a protein, circular strands of DNA, a virus or several types of polymers as self-assembly agents. 1: dendrimer (repetitively branched polymer), 2 and 3: linear polymers leading to flat or spherical networks depending on the solvent or the surface (adapted from Ofir et al. [OFI 08]). For a color version of this figure see www.iste.co.uk/lacaze/nano.zip*

After adding regularly spaced out "anchor points" onto the strands of DNA, each one made up of a segment of another DNA strand, it is possible to assemble a well-structured, regular network of receptor sites. Each receptor site (fragment of a DNA strand) can associate with a complementary strand of DNA, on which an NP will have been fixed.

The NP is spontaneously positioned in the network as a result of a hybridization reaction between two fragments of the DNA strands, leading to parallel lines of NPs (Figure 3.17).

The technique is amazing and using the very unusual chemistry of DNA, based on the association of complementary strands, provides interesting perspectives for designing networks. However, this is a very sophisticated method, still a long way from large-scale applications.

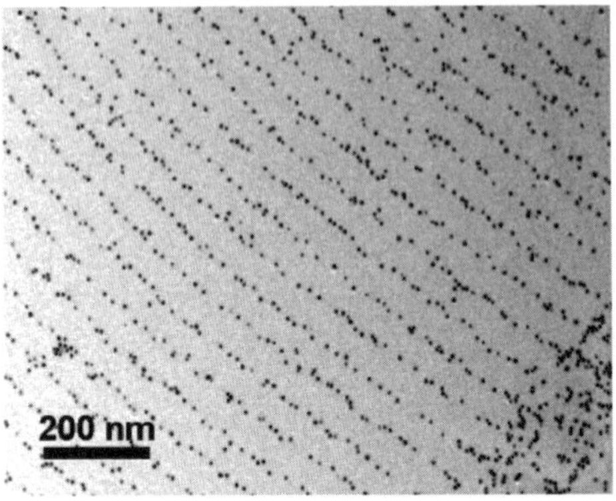

Figure 3.17. *TEM image of a network of gold NPs obtained via a grid of DNA strands arranged on a surface, on which the gold NPs are attached. The spacing between the rows of NPs is equal and approximately 62.9 ± 0.8 nm; the size of the NPs is 6.2 ± 0.8 nm and the average distance between particles of the same row is approximately 12 ± 3 nm (adapted from Le et al. [LE 04])*

3.2.2. *Assembling by ultrathin alumina membranes*

Using alumina porous membranes to insert chemical compounds into pores is a rather old electrochemical procedure, having been described in detail in the 1970s.

The procedure has since improved considerably creating a new interest in this field of research, explained by all the new possible applications in the field of surface nanostructuring and obtaining organized networks of nano-objects.

Figure 3.18(a) shows an image of an alumina membrane obtained electrochemically and after a double electrolysis, carried out on a thin layer of aluminum of only several hundred nanometers thick.

The membrane appears to be structured throughout its entire depth. It is composed of perfectly parallel cylindrical tubes, regularly spaced in relation to another. Their diameter, between 100 and 200 nm, can be easily adjusted by carefully choosing the operating conditions (electrolysis voltage, choice of electrolyte and reaction time).

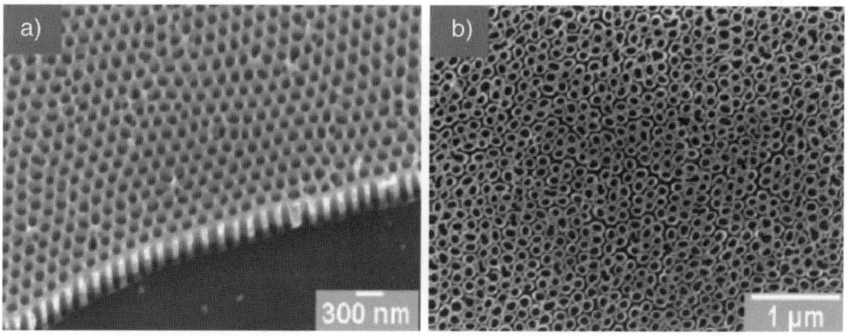

Figure 3.18. *a) SEM image of an alumina membrane of approximately 300 nm thick obtained by electrolysis and attached onto a silicon base; b) tubes of titanium dioxide obtained from the previous alumina membrane (adapted from Tan et al. [TAN 08])*

These membranes can subsequently be used as masks during the deposition of other materials, such as a regular network of tubes of titanium dioxide (TiO_2) obtained from the previous alumina mask, after the alumina matrix is dissolved with chemicals (Figure 3.18(b)).

3.2.3. *Assembling by colloidal lithography*

The principle of this method is simple. One or several monolayers of polystyrene or silica nanospheres suspended in a liquid (colloidal solution of particles) are deposited onto a surface. As the liquid evaporates, this colloidal layer of nanospheres spontaneously organizes itself into a compact hexagonal network forming the surface mask.

Therefore, a regular network of crevices is created, in which other particles may be inserted. This technique, developed in the 1980s, is often called "colloidal lithography". The corresponding networks are composed of colloidal crystal monolayers (CCMs) (Figure 3.18).

Organizing nanospheres into compact monolayers may be favored by using several variants (evaporation of the solvent on a turning plate, adsorption of particles by vertical dipping of the surface when the particles are arranged on the liquid–gas interface, deposition assisted by an electric field in the case where the particles are electrically charged).

The regularity of the CCM is extraordinary, as demonstrated in the SEM images obtained in the case of compact or non-compact networks of silica spheres (Figure 3.20).

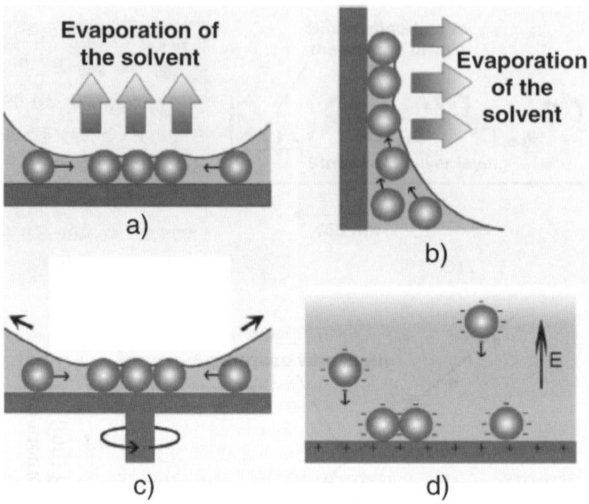

Figure 3.19. *Some examples of strategies used to create monolayers of tiny spheres regularly arranged on a surface: a) Natural evaporation of a solvent; b) vertical dipping of the surface in a colloidal solution; c) induced evaporation by rotating the surface (spin-coating); d) deposition assisted by an electric field, in the case of electrically charged spheres (adapted from Ye and Qi [YE 11])*

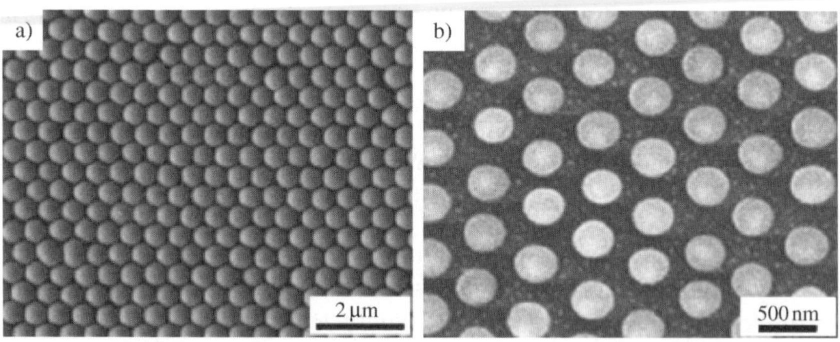

Figure 3.20. *SEM images of monolayers of colloidal spheres of silica assembled on a solid surface: a) Compact network obtained by dipping of the surface (dip-coating) in a colloidal solution of silica spheres (560 nm in diameter), adapted from Li et al. [LI 10]; b) non-compact network obtained by the evaporation of a solvent and rotation of the surface (spin-coating) (adapted from Khanh et al. [KHA 09])*

From these networks of nano- or microspheres (polystyrene or silica), it is possible to structure the surface in a diverse number of ways; this is because the interstitial dimensions can be adjusted by selecting sphere diameters between 50 nm and 4–5 μm. The possible modifications that can potentially be made to these surfaces are immense; this allows all sorts of structures to be obtained and their innumerable applications can be predicted.

Numerous variations in this technique have been described.

Evaporating a metal on top of the mask of nanospheres, then removing the spheres by simple dissolution (with the aid of a solvent, in the case of polystyrene) is a very widespread technique. It allows metallic nanoplots networks to be obtained, arranged in regular hexagons, similar to honeycomb structures (Figure 3.21(a), (d)).

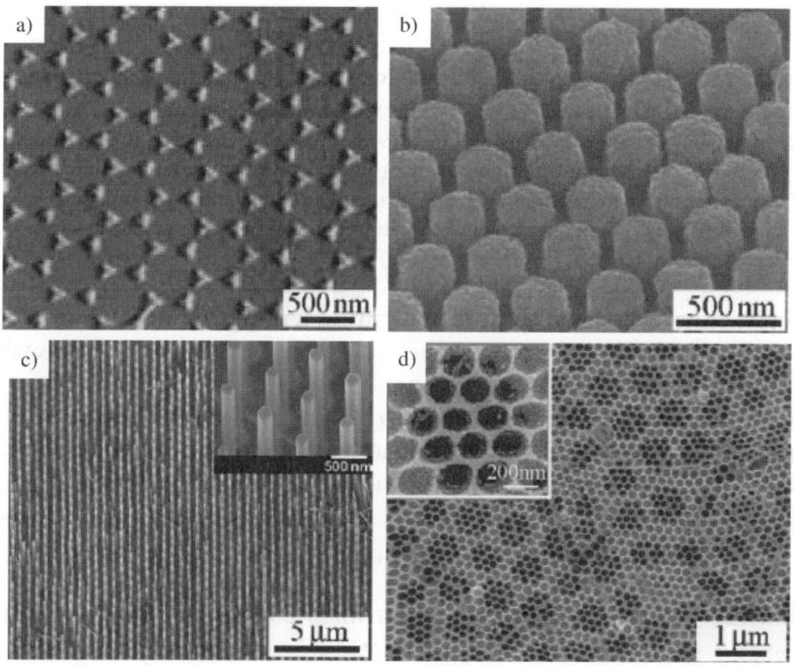

Figure 3.21. *Examples of different surface structures achieved using diverse techniques and via CCM: a) A network of silver nanoplots deposited by evaporation in a vacuum, adapted from Haynes et al. [HAY 01]; b) nanocylinders of titanium dioxide (TiO_2) formed by laser impulses, adapted from Li et al. [LI 09]; c) nanocylinders of zinc oxide (ZnO) obtained by evaporation in a vacuum, adapted from Zhang et al. [ZHA 10]; d) porous membrane of indium oxide, adapted from Jia et al. [JIA 09]*

Conversely, carrying out "dry" ionic etching in the gaps that are not protected by the spheres (bombarding the surface with ions under vacuum conditions) allows regularly spaced nanopillars to be formed (Figure 3.21 (b), (c)). Finally, via electrochemical processes, it is also possible to structure a surface initially modified by a CCM by depositing diverse materials into the crevices via electrolysis, for example conducting polymers onto different metals (Santos *et al.* [SAN 10]).

Because of its low cost, its simplicity and its great reproducibility, and due to the fact that it can be applied onto large surfaces, colloidal lithography undoubtedly involves a new and original surface nanostructuring method, capable of large-scale applications.

3.3. Conclusions

A large number of techniques are now available to develop nanomaterials with extremely varied shapes and dimensions. After discovering gold colloidal solutions, which gave rise to an impressive number of applications, a large number of colloidal solutions of very diverse materials (metals, SC or polymers) could be produced. Based primarily on precipitation reactions of chemicals in solution, it should however be noted that these colloidal solutions are composed of NPs that are still encased in organic molecules, vital for their preservation. And they are partly responsible for the properties of the NPs.

Organizing these NPs or diverse nanomaterials into networks on surfaces is also an important objective. It can be done by self-assembly of NPs in solution or even by the combined use of masks and lithographic techniques. This is a rapidly expanding area of research, with great interest in the numerous possible applications it offers (see Chapter 4).

3.4. Bibliography

[BAT 99] BATES F.S., FREDRICKSON G.H., "Block copolymers: designer soft materials", *Physics Today*, vol. 52, pp. 32–38, 1999.

[DAN 04] DANIEL M.-C., ASTRUC D., "Gold nanoparticles: assembly, supramolecular chemistry, quantum-size-related properties, and applications toward Biology, catalysis, and nanotechnology", *Chemical Reviews*, vol. 104, pp. 293–346, 2004.

[GEO 10] GEORGELIN T., MAURICE V., MALEZIEUX B., SIAUGUE J.-M., CABUIL V., "Design of multifunctionnalized γ-Fe2O3@SiO2 core-shell nanoparticles for enzymes immobilization", *Journal of Nanoparticle Research*, vol. 12, pp. 675–680, 2010.

[HAY 01] HAYNES C.L., VAN DUYNE R.P., "Nanosphere lithography: a versatile nanofabrication tool for studies of size-dependent nanoparticle optics", *Journal of Physical Chemistry B*, vol. 105, pp. 5599–5611, 2001.

[JIA 09] JIA L., CAI W., WANG H., SUN F., LI Y., "Hetero-apertured micro/nanostructured ordered porous array: layer-by-layered construction and structure-induced sensing parameter controllability", *ACS Nano*, vol. 3, pp. 2697–2705, 2009.

[KHA 09] KHANH N.N., YOON K.B., "Facile organization of colloidal particles into large, perfect one and two-dimensional arrays by dry manual assembly on patterned substrates", *Journal of the American Chemical Society*, vol. 131, pp. 14228–1423, 2009.

[KIM 10] KIM H.-C., PARK S.-M., HINSBERG W.D., "Block copolymer based nanostructures: materials, processes, and applications to electronics", *Chemical Reviews*, vol. 110, pp. 146–177, 2010.

[LE 04] LE J.D., PINTO Y., SEEMAN N.C., MUSIER-FORSYTH K., TATON T.A., KIEHL R.A., "DNA-templated self-assembly of metallic nanocomponent arrays on a surface", *Nano Letters*, vol. 4, pp. 2343–2347, 2004.

[LEI 11] LEI Y., YANG S., WU M., WILDE G., "Surface patterning using templates: concept, properties and device applications", *Chemical Society Reviews*, vol. 40, pp. 1247–1258, 2011.

[LI 09] LI Y., FANG X., KOSHIZAKI N., SASAKI T., LI L., GAO S., SHIMIZU Y., BANDO Y., GOLBERG D., "Periodic TiO2 nanorod arrays with hexagonal nonclose-packed arrangements: excellent field emitters by parameter optimization", *Advanced Functional Materials*, vol. 19, pp. 2467–2473, 2009.

[MAD 08] MADUENO R., RAÏSÄNEN M.T., SILIEN C., BUCK M., "Functionalizing hydrogen-bonded surface networks with self-assembled monolayers", *Nature*, vol. 454, pp. 618–621, 2008.

[MAS 81] MASSART R., "Preparation of aqueous magnetic liquids in alkaline and acidic media", *IEEE Transactions on Magnetics*, vol. 17, pp. 1247–1248, 1981.

[MEL 03] MELOSH N.A., BOUKAI A., DIANA F., GERARDOT B., BADOLATO A., PETROFF P.M., HEATH J.R., "Ultra high density nanowire lattices and circuits", *Science*, vol. 300, pp. 112 115, 2003.

[MIC 05] MICHALET X., PINAUD F.F., BENTOLILA L.A., TSAY J.M., DOOSE S., LI J.J., SUNDARESAN G., WU A.M., GAMBHIR S.S., WEISS S., "Quantum dots for live cells, in vivo imaging, and diagnostics", *Science*, vol. 307, pp. 538–544, 2005.

[OFI 08] OFIR Y., SAMANTA B., ROTELLO V.M., "Polymer and biopolymer mediated self-assembly of gold nanoparticles", *Chemical Society Reviews*, vol. 37, pp. 1814–1825, 2008.

[OGL 11] OGLI A.S., "Nanobio applications of quantum dots in cancer: imaging, sensing and targeting", *Cancer Nano*, vol. 2, pp. 1–19, 2011.

[ROG 11] ROGACH A.L., "Fluorescence energy transfer in hybrid structures of semiconductor nanocrystals", *Nano Today*, vol. 6, pp. 355–365, 2011.

[SAN 10] SANTOS L., MARTIN P., GHILANE J. LACAZE P.-C, RANDRIAMAHAZAKA H., ABRANTES L.M., LACROIX J.-C., "Electrosynthesis of well-organized nanoporous poly(3,4-ethylenedioxythiophene) by nanosphere lithography", *Electrochemistry Communications*, vol. 12, pp. 872–875, 2010.

[SCH 08] SCHMID G., "The relevance of shape and size of Au 55 clusters", *Chemical Society Reviews*, vol. 37, pp. 1909–1930, 2008.

[TAN 08] TAN L.K., CHONG M.A.S., GAO H., "Free-standing porous anodic alumina templates for atomic layer deposition of highly ordered TiO2 nanotube arrays on various substrates", *Journal of Physical Chemistry C*, vol. 112, pp. 69–73, 2008.

[WAN 10] WANG L., LUO J., SCHADT M.J., ZHONG C.-J., "Thin film assemblies of molecularly-linked metal nanoparticles and multifunctional properties", *Langmuir*, vol. 26, pp. 618–632, 2010.

[WIN 98] WINFREE E., LIU F., WENZLER L.A., SEEMAN N.C., "Design and self-assembly of two-dimensional DNA crystals", *Nature*, vol. 394, pp. 539–544, 1998. (A good description of the technique is also given on Wikipedia – DNA Origami)

[YAM 11] YAMAN M., KHUDIYEV T., OZGUR E., KANIK M., AKTAS O., OZGURI E.O., DENIZ H., KORKUT E., BAYINDIR M. "Arrays of indefinitely long uniform nanowires and nanotubes", *Nature Materials*, vol. 10, pp. 494–501, 2011.

[YAN 10] YANG P., YAN R., FARDY M., "Semiconductor nanowire: what's next?" *Nanoletters*, vol. 10, pp. 1529–1536, 2010.

[YE 11] YE X., QI L., Revue. "Two-dimensionally patterned nanostructures based on monolayer colloidal crystals: controllable fabrication, assembly, and applications", *Nano Today*, vol. 6, pp. 608–631, 2011.

[ZHA 10] ZHANG J., LI Y., ZHANG X., YANG B., "Colloidal self-assembly meets nanofabrication: from two-dimensional colloidal crystals to nanostructure arrays", *Advanced Materials*, vol. 22, pp. 4249–4269, 2010.

Chapter 4

Some Amazing Properties of Nanomaterials and of Their Assembly into Networks

Owing to the confinement of matter, the physicochemical properties of a material are considerably modified with regard to its solid state.

There are essentially two causes of these modifications in properties: on the one hand, an entirely geometric effect, which increases the number of atoms on the surface in relation to the inner atoms and hence is going to modify the catalytic properties of the material. On the other hand, a modification in the energetic characteristics of the material, due to the quantum effect; this is apparent as soon as the size of objects becomes less than several nanometers.

Particularly in the case of SCs and metals, significant variations in electron energy levels are observed; these are at the root of the new optoelectronic properties, completely different to the material when it is in its solid state.

4.1. The first effect of the confinement of matter: unusual catalytic and physicochemical properties

Increasing reaction times is a major problem, frequently seen in the chemical industry, causing multiple problems regarding manufacture[1].

1 A chemical reaction is a process through which one or several compounds (reactive) are transformed into a new product. This process is rather quick; its reaction time (amount of matter transformed per second) depends on the experimental conditions (concentration and temperature).

Numerous reactions that commence slowly (e.g. hydrogenation of hydrocarbons such as benzene or ethylene, regularly carried out in the industry) can be sped up when the reagents come into contact with solid compounds, which act as catalysts.

It is the heterogeneous contact (solid–gas or solid–liquid) between catalyst and reactants (in the gaseous or liquid state) that accelerates the reaction, and hence the transformation of reactants into products is quicker. Obviously, this catalytic effect occurs via the surface atoms, and therefore the more divided the material is, the larger the number of atoms there will be on the surface with regard to inner atoms.

In fact, a simple calculation, taking the average dimensions of atoms into account, shows that the ratio between the number of atoms on the surface and the number of atoms in the core rapidly increases when the size of the particle decreases.

So, with a gold metallic cube of micrometric dimensions (1 μm^3 or a billionth of a millimeter cubed), this ratio is equal to 0.0004, and increases to 0.6 when the volume is 1 nm^3, that is a billion times smaller.

As it is the atoms on the surface that intervene in the catalytic process, we can appreciate that the extremely divided materials will be better catalysts than the same material in the solid state.

For example, ultrathin gold NPs (less than 1 nm) deposited on titanium dioxide (TiO_2) transforms, at room temperature and in the presence of oxygen and carbon monoxide (CO), which is extremely poisonous, into carbon dioxide (CO_2), which is much less toxic. The catalytic effect is even more remarkable than the same reaction carried out on solid gold[2] in that does not give rise to any transformation; this verifies the particular role these surface atoms play, with regard to the core atoms [CHE 08].

4.2. The optoelectronic properties of NPs due to confinement

Without going into detail about quantum physics, some basic concepts are in fact useful in order to better understand the properties of electrons

2 The dominant properties of the surface atoms with regard to the inner atoms disappear as soon as the size of particles becomes greater than 10 nm.

contained in materials, whether metals, SCs or insulators, in the solid state or in the form of NPs.

4.2.1. *Some concepts of physics that can be applied to solid materials*

Applying quantum mechanics to solid crystal materials indicates that the electrons are distributed according to increasing energy levels (corresponding to quantum states) and that there cannot be more than two electrons in each energy level (the Fermi principle).

The important thing is that, considering the high densities of these quantum states (due to a very large number of atoms per cm^3), these levels are extremely close to one another. It is no longer possible to distinguish between them, leading to the formation of energy bands in which the electrons are distributed.

In the case of insulating and semiconducting materials, the electrons are distributed across two distinct bands, separated by a band gap E_g, which corresponds to a so-called forbidden energy band, whose width is a feature of the material, and in which no electrons exist.

The lowest energy band is called the valence band (VB), and is completely occupied by electrons; the highest band of energy, the conduction band (CB), can be completely emptied (in the case of an insulator) or partially occupied by electrons (in the case of an SC).

The electrons situated in the VB have no way of moving inside this band (all the energy levels are occupied by electrons); they cannot even "jump" into the CB to occupy vacant energy levels (thermal energy that electrons can gain is too weak for them to be able to cross the energy barrier and move into the CB).

The calculation indicates that the density of charge carriers (electrons found in the CB) is very low at, approximately $10^{-27}/cm^3$, almost 0. This explains why insulating materials such as diamond do not conduct electricity[3] (Figure 4.1(a)).

---

3 Diamond is known to have an extremely high E_g value (~5.5 eV), whereas silicon, the most frequently used semiconductor, has an E_g value five times smaller (1.1 eV) than that of diamond.

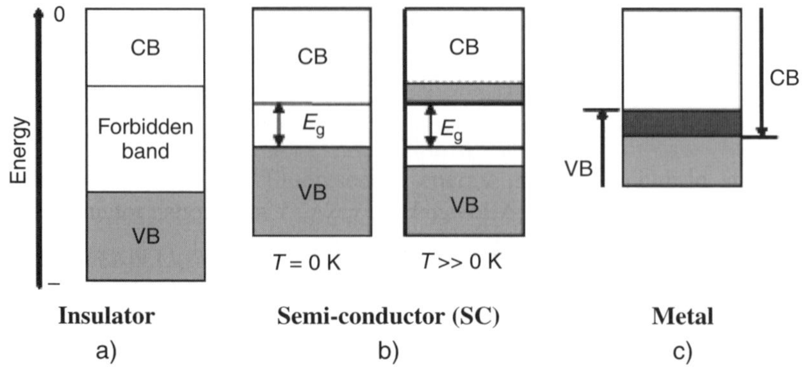

Figure 4.1. *Simplified diagrams showing the layout of the valence band (VB) and conduction band (CB) according to the energy E for an insulator, a semiconductor (SC) and a metal (0 energy corresponds to the energy of an electron in a vacuum). The areas occupied by electrons are indicated in gray: a) in the case of an insulator: the forbidden bandwidth is very large; b) in the case of a SC: the forbidden bandwidth (E_g) is much smaller than for an insulator. At 0 degree Kelvin (0 K or –273 °C) the VB is completely occupied by electrons; for a temperature much greater than 0 K (T >> 0 K), the electrons gain enough energy to jump into the CB. These electrons are able to move about under the effect of an electric field, which gives the material semiconducting properties; c) in the case of a metal: the VB and the CB "overlap" (in black, the area shared by the VB and the CB). Whatever the temperature, a large number of electrons are found in the conduction band; the metal is an extremely good conductor*

With SCs, the energy diagram is practically the same as that of insulators, with the only difference being a smaller forbidden bandwidth (Figure 4.1(b)). Consequently, the CB becomes accessible to the electrons in the VB.

At absolute zero (0 K or –273 °C), the CB is still unoccupied, but at a higher temperature (T >> 0 °K), a small number of electrons from the VB acquire enough energy (thermal) "to jump" into the CB, which possesses a large number of unoccupied levels.

By jumping into the CB, they create holes in the VB (vacant quantum levels corresponding to the absence of electrons); this also provides new opportunities for the valence electrons to move about.

The holes and the electrons (situated in the VB and CB, respectively), formed in equal amounts, are SC charge carriers. They have sufficient numbers ($10^{19}/cm^3$ or 10 trillion holes and electrons per cm^3) to give the SC properties of a weak conductor (Figure 4.1(b)).

In the case of metals, the CB and the VB partially overlap (Figure 4.1(c)). Consequently, this allows the electrons in the VB to move about freely in the CB, which has a large number of unoccupied energy states.

The density of charge carriers (free or mobile electrons) is much higher, approximately $10^{23}/cm^3$, that is 10,000 times greater than that of SCs. This is the reason why metals are excellent conductors of electricity.

4.2.2. *The plasmon resonance effect and the optical properties of gold NPs*

The interaction of light with a metallic surface is the cause of a very specific effect known as plasmon resonance (PR). It is responsible for the color of metals and the effect is particularly intense with gold and silver.

The explanation behind this phenomenon is found in the properties of the surface electrons, very mobile and similar to a gas of electrons, capable of oscillating collectively at a certain frequency that is specific to each metal.

Monochromatic light radiation, characterized by a wavelength λ[4] and by an oscillation frequency ν, can interact with electrons and cause a collective oscillation of electrons, when the oscillation frequencies of light and of the electrons are close to each other.

It is said that there is resonance, and the phenomenon is accompanied by a rather intense absorption of light.

Visible light has an infinite number of light radiations of distinct wavelengths. One of them has the same oscillation frequency as the electrons and will strongly interact with the electrons in the metal in a synchronous fashion. The phenomenon is accompanied by a transfer of energy between light radiation and the oscillating electrons.

For a specific frequency, this results in the metal absorbing the electromagnetic radiation, which induces a particular color, yellow in the case of solid gold, gray in the case of silver. This effect is known as surface plasmon resonance (SPR).

---

4 Visible light can be considered as the overlapping of monochromatic light waves, each one corresponding to a characteristic wavelength λ and to a frequency ν such as $\nu = c/\lambda$ (c: speed of light).

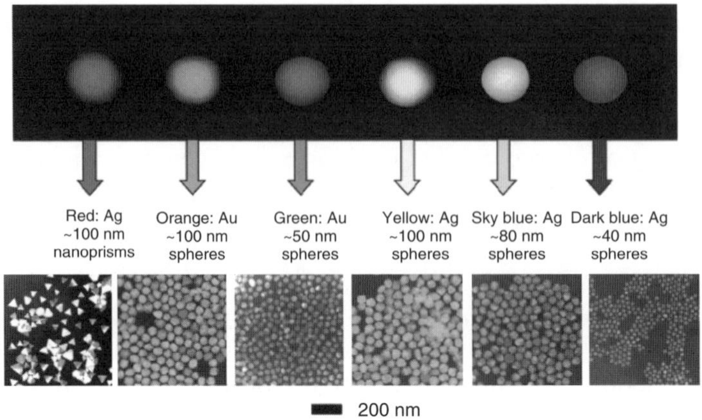

Figure 4.2. *Colors of gold (Au) and silver (Ag) NPs in various forms. Note the very different colors between the 100 nm Ag nanoprisms and 100 nm Ag spheres, red and yellow, respectively, which highlight how important is the shape to the plasmon resonance (adapted from Rosi et al. [ROS 05]). For a color version of this figure see www.iste.co.uk/lacaze/nano.zip*

When a metal is found in the form of an NP, the oscillation frequency of electrons of a particle is modified with regard to that of the solid metal. The effect is localized on the surface of the particle, and the resonance of the surface electrons therefore depends on the size of the NPs. This is called localized surface plasmon resonance (LSPR).

This means that every particle size will have a specific color corresponding to it; this is also dependent on the shape of the NP (Figure 4.2).

So, with gold NPs of 6 nm in diameter, the plasmonic resonance (PR) occurs at wavelengths between 650 and 700 nm (corresponding to absorption of red light). Then the light reflected will be blue-violet (complementary to the absorbed red color)[5]. When the size of the particle increases, the plasmonic oscillation frequency decreases, the absorption of light occurs in the red part of the spectrum, and therefore the color reflected by the particle shifts toward blue-green (for NPs of 50 nm, they are green in color, Figure 4.2).

5 The visible spectrum of light extends over a range of wavelengths from 380 nm (violet) to 760 nm (red). White light can be considered as the overlapping of blue, green and red lights; this corresponds to the fact that the eye, sensitive to these three colors, recreates white light by combining these three colors. When light is absorbed in the blue-green spectrum (between 500 and 550 nm), the color of the NP appears red (by subtraction). Conversely, if the light is absorbed in the red part of the spectrum (between 650 and 700 nm), the color will be blue-green.

The change in color linked to the size of the NPs also depends on aggregation. This effect was used to create biosensors capable of revealing certain biological molecules to the naked eye.

A colloidal solution of gold NPs can in fact give rise to significant changes in color when the gold NPs, initially isolated from one another, begin to form aggregates of several NPs.

So, a solution containing gold NPs of approximately 20–30 nm in diameter, red in color, becomes bluish when the same NPs associate in pairs or in groups of three or four NPs. This color change phenomenon related to the aggregation of NPs has been successfully used to detect strands of DNA in solution (Box 3.3, DNA).

This is a very clever procedure involving the assembly (hybridization) and separation of strands of DNA attached to gold NPs (Figure 4.3).

This is done using two sets of gold NPs, approximately 15 nm in diameter, to which fragments of DNA **a′** and **b′** are attached, the respective complementary halves of **a** and **b** of the DNA strand **a–b** that is to be detected in solution.

When there is no **a–b** DNA strand in solution, the gold NPs are separate giving the solution a red color. When DNA is present, the gold NP-**a′** and NP-**b′** nanoparticles will bind to one another via the two halves a and b of the **a–b** strand and form aggregates [NP-**a'**-(**ab**)-**b'**-NP] which causes the color of the solution to change from red to blue (Figure 4.3).

Another feature is that the hybridization phenomenon between the complementary DNA strands can only occur below a critical temperature T_m. When $T < T_m$, hybridization occurs and when $T > T_m$, the strands separate.

4.2.3. *Surface enhanced Raman scattering*

When light interacts with gold NPs, color is not the only effect that is produced. Besides PR, which leads to a partial absorption of light and to its diffusion (reflection in all directions), an electric field, localized on the interface, is also produced and oscillates at the same frequency as that of the PR.

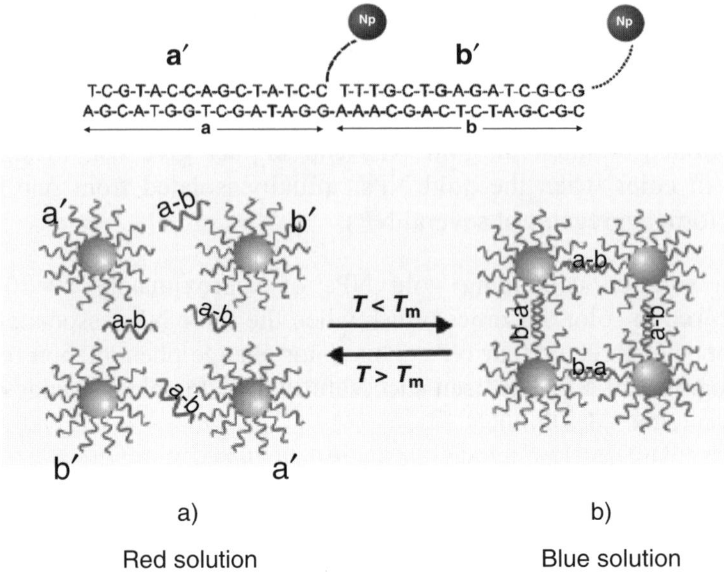

Figure 4.3. *The DNA strand that we want to detect is made up of the chain* ***a–b****. The gold NPs represented by spheres bring the strands* ***a′ (NP-a′)*** *or* ***b′(NP-b′)*** *to the surface, complementary to* ***a*** *and* ***b****. a) When the temperature T is greater than T_m, there is no hybridization between* ***a*** *and* ***a′****,* ***b*** *and* ***b′****: the NPs are separated from one another; this corresponds to a red-colored solution. b) On cooling the solution ($T < T_m$) hybridization occurs (association of* ***a*** *and* ***a′****,* ***b*** *and* ***b′*** *by the paring up of bases A, T, C and G. The gold NPs clump together by forming aggregates* ***NP-a′-(ab)-b′-NP*** *and the solution turns blue (adapted from Elghanian et al. [ELG 97]). For a color version of this figure see www.iste.co.uk/lacaze/nano.zip*

This electrical field immediately next to the surface is very strong, and rapidly weakens as the distance to the NP increases. It can interact with molecules adsorbed onto the surface of the NP. This leads to the scattering of light, characteristic of the molecule, known as Raman scattering[6].

This effect is particularly intense with silver NPs and gives rise to important analytic applications, which allow extremely small amounts of chemical compounds to be detected.

---

6 Generally, this Raman scattering has a low intensity, and it was only much later that various metals such as gold and especially silver NPs were found to be capable of producing a strong enhancement of this scattering phenomenon. Localized close to the surface of the metal, it is appropriately called SERS or Surface Enhanced Raman Scattering).

4.2.4. *The photothermic effect or how to heat up gold NPs*

In general, the movement of electrons within a material results in an energy loss in the form of heat. This heat energy produced is much greater when the electrical resistance of the material is higher (as opposed to the spontaneous movement of electrons). In the case of gold, the electrical resistance is very low and the electrical current passing through solid gold does not produce any significant heat.

This is no longer the case with NPs of small dimensions. The oscillating electrons in the gold NP always give off less heat, but due to being confined to a very small volume, it is going to cause an increase in temperature that is inversely proportional to the volume of the particle.

For NPs of only a few nanometers in diameter, the absorption of light corresponding to the PR frequency of the NP results in a significant increase in temperature (approximately 20–30 °C).

This local warming is sufficient to destroy biological cells that would have been targeted by gold NPs; it is used particularly in the photothermal treatment *in vivo* of cancerous cells.

4.2.5. *The optoelectronic properties of Quantum Dots*

The energetic characteristics of VB and CB, established in the case of solid semiconductor materials, are greatly modified when the dimensions of the material decrease to sizes of only a few nanometers.

The reduced number of electrons causes the energetic structure of the nanomaterial to resemble that of a group of atoms, and when the size becomes extremely small (approximately 1 nm) the energy bands disappear and give rise to distinct levels. Metals become Nanodots and SCs are therefore Quantum Dots (QDots) (Figure 4.4).

This difference in energy (gap), between the highest occupied energy state and the lowest vacant level, is the underlying cause of numerous optoelectronic properties and, in particular, it determines the color of the fluorescence emitted by the semiconducting NPs (Box 3.1, Luminescence).

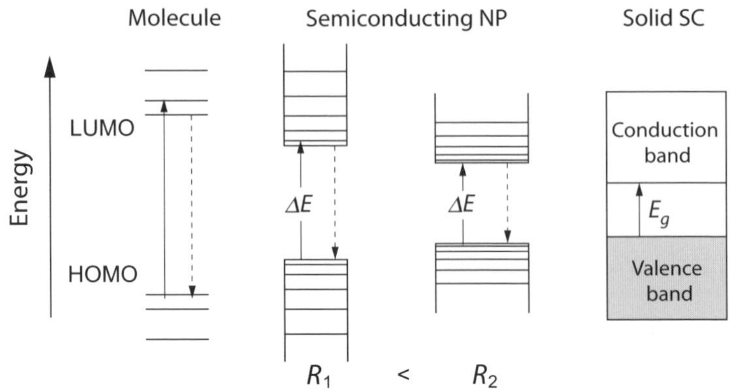

Figure 4.4. *Evolution of the energy "gap" as we pass from a solid semiconductor to semiconducting nanocrystals. As the R (radius) of the nanocrystal decreases ($R_1 < R_2$), the band structure disappears and gives rise to discrete energy bands (bandwidth of 0). The arrow pointing upward represents the jump in energy that the electron must make in order to pass from an occupied level to a vacant level (the excited state of the electron), whereas the downward pointing arrow represents the electron returning to its initial state, accompanied by the emission of photons or heat given off. In the case of a molecule, the highest level occupied by electrons is the highest occupied molecular orbital (HOMO) and the lowest vacant level is the lowest unoccupied molecular orbital (LUMO)*

Therefore, semiconducting NPs of cadmium telluride (CdTe) emit a green, a yellow or a red light, respectively, as their size increases from 2 to 6 nm (Figure 4.5).

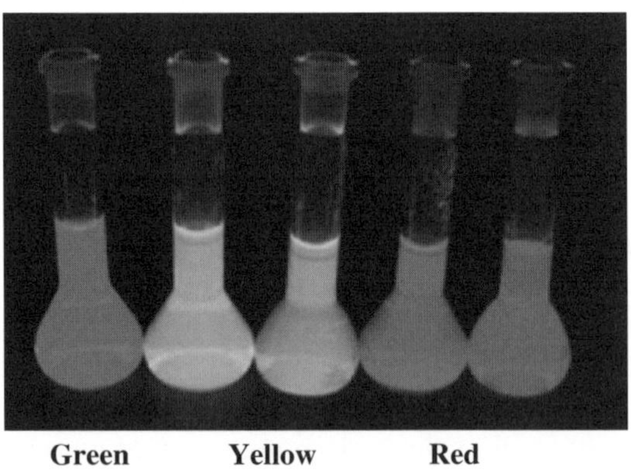

Figure 4.5. *Light emitted by colloidal solutions of CdTe (Cadmium telluride) NPs lit up under UV light. From left to right the size of the NPs increases from 2 to 6 nm and the light emitted shifts from green to yellow and red (adapted from Rogach [ROG 11]). For a color version of this figure see www.iste.co.uk/lacaze/nano.zip*

The fact that semiconducting NPs can be used as an optical marker is currently the subject of applied research throughout the field of medicine; as we will see later on their possible detection at very low concentrations makes them preferred tools when developing new *in vivo* diagnostic techniques.

4.3. The amazing properties of NP networks or nanostructured surfaces

Arranging NPs into networks produces significant modifications in the properties of the NPs, mainly due to the fact that we are no longer dealing with separate phenomena, but instead combined effects, effects that can also be observed in the case of nanostructured surfaces.

A certain number of properties (surface wettability, reflection of light off a micro- or nanostructured surface, electron-emissive surfaces under the effect of an electrical field, surface enhanced Raman scattering (SERS) properties of metallic nanodot networks) are all directly linked to the periodicity of the network (spacing between nano-objects), to the nature and to the morphology of nanomaterials on a surface.

4.3.1. *Wettability of structured surfaces*

This is an important property that at the same time depends on the chemical nature and the roughness of the material's surface. This property can be directly observed when a drop of water is placed on the flat surface of a material.

The material is said to have a large wettability if the water droplet spreads perfectly over the surface in the form of a homogeneous transparent film. Conversely, if the drop of water "rolls about" over the surface, the wettability is negligible (the surface is said to be hydrophobic).

This property can be quantitatively assessed by measuring the contact angle θ between the surface and the water droplet (Figure 4.6).

These wettability and non-wettability properties of surfaces are at the heart of important practical applications such as self-cleaning of surfaces, anti-fog glass, the adhesion and non-adhesion of microorganisms onto surfaces, and also the protection of metallic surfaces against corrosion.

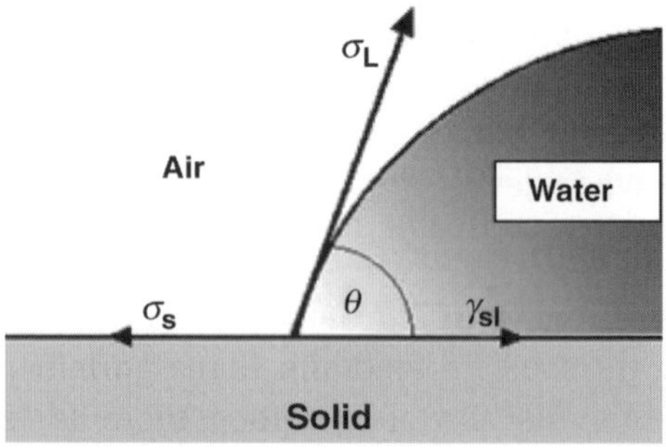

Figure 4.6. *Characterizing the surface wettability. When the contact angle θ is equal to 0°, the surface is considered to have maximum wettability, whereas the opposite occurs when θ is close to 180°. σ_s and σ_L are the superficial energies of the solid and the liquid, respectively; γ_{sL} is the surface tension between the water and the solid; σ_s, σ_L and γ_{sL} are variables calculated from θ. A hydrophobic or a hydrophilic surface is characterized by a weak or a strong γ_{sL} value, respectively*

The initial wetting or non-wetting phenomenon only occurs if atoms on the surface have hydrophilic or hydrophobic properties. Usually, this hydrophilic or hydrophobic characteristic is given to the material when molecules that are themselves hydrophilic or hydrophobic bond onto its surface (Box 4.1).

These properties (of a chemical origin) can then be considerably amplified when a micro- or a nanostructuration of the surface is created. These structural effects are particularly significant when the holes and peaks created in the surface are related to droplet size.

Superhydrophobic surfaces can therefore be obtained (contact angles greater than 150°, instead of 80–90° for a flat surface). This is the Lotus effect, which can also be obtained by artificially dispersing micrometric asperities over a hydrophobic surface (Figure 4.7).

In the same way, using hydrophilic surfaces characterized by contact angles of approximately 20–30°, it is possible to make these surfaces superhydrophilic (with contact angles close to 0°) by creating the same type of structuring, and in this case the effect can be used to create anti-fog glass[7].

---

7 The appearance of fog on a glass surface corresponds to the formation of micro-droplets, which remain transparent as long as their diameter is less than half of the smallest wavelength (380 nm) of the visible spectrum of light; they become opaque as soon as their size goes beyond this limit. The appearance of fog can be prevented by making the surface superhydrophilic; this leads to a transparent homogenous film of water.

Box 4.1. Hydrophobic–hydrophilic

Hydrophobic. This is a characteristic property of materials such as wax or Teflon. When water comes into contact with these surfaces, it does not spread out, but instead forms tiny spherical drops on the surface, which tend to roll about in whichever way the material is tilted. This property is linked to the chemical structure of molecules, formed from long non-polar hydrocarbon (wax) or fluorocarbon (Teflon) chains, with completely total absence of localized electric charge. This eliminates the possibility of hydrogen bonds forming with water molecules (polar) and makes compounds insoluble in water. They are, however, very soluble in oils and organic solvents (liquid hydrocarbons), which are also hydrophobic; this is why these compounds can sometimes be called oleophilic compounds. The contact angle of a water droplet with a hydrophobic surface is greater than 90°.

Hydrophilic. This property is the opposite of hydrophobicity and unlike the previous, it concerns all materials and chemical compounds with polar groups (carriers of electric charge), allowing them to form hydrogen bonds with water molecules. It is usually chemical compounds with OH groups (alcohols, acids), C = O (ketones) that form hydrogen bonds with water molecules. They are soluble in water and insoluble in non-polar organic solvents. One drop of water placed onto a hydrophilic surface tends to spread out and form a contact angle less than 40°.

Amphiphilic compounds. These molecules are both hydrophobic and hydrophilic at the same time. They are usually composed of a long hydrophobic carbonaceous chain and a polar hydrophilic carboxylate head ($-COO^-Na^+$), or a sulfonate head ($-SO_3^-Na^+$).

This duality gives them very specific properties: they are soluble in both water and oils.

The most common are spherical assemblies corresponding to direct micelles, whose polar heads are in contact with water and the hydrophobic chains (tails) crowd together to prevent any contact with water. Their size varies between several nanometers to approximately 50 nm.

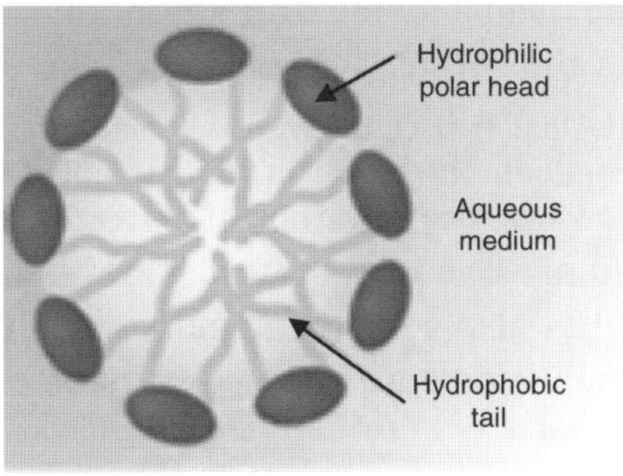

Soaps are made up of amphiphilic molecules (fatty acids + sodium hydroxide). They are capable of dissolving oily compounds and at the same time being soluble in water, which makes them good detergents.

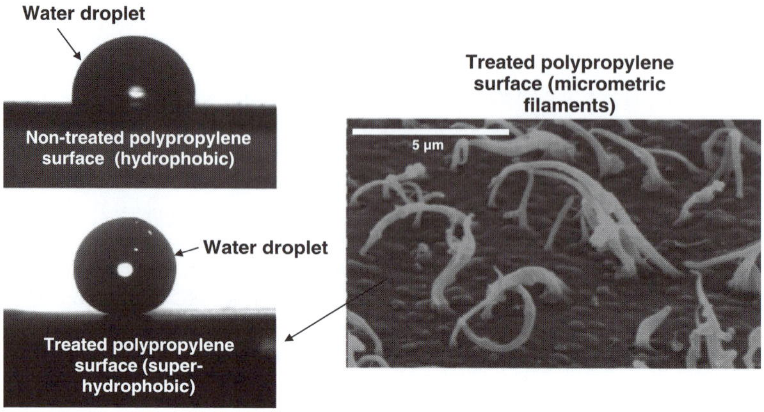

Figure 4.7. *Example of the evolution of a flat hydrophobic polypropylene surface toward a superhydrophobic state after structuring the surface with micrometric polypropylene filaments. The contact angle between the drop of water and the surface varies from 90° in the case of a non-treated flat surface (hydrophobic) to more than 150° in the case of a microstructured surface (superhydrophobic) (adapted from Hsu and Sigmund [HSU 10])*

The superhydrophobicity of a surface is rather interesting in practice. It is being increasingly used in construction when making self-cleaning facades, and also in the photo-solar domain for protecting photovoltaic panels against the accumulation of dust.

The self-cleaning principle is easy to understand. The surface must have a significant inclination angle and a high contact angle with water (approximately 140–150°); this implies a premicrostructuring of the hydrophobic surface. This microstructuring leads to the formation of small droplets of water, which by rolling over the surface will get rid of the dust particles by leading them off the surface (Figure 4.8).

Solar panels based on treated amorphous or polycrystalline silicon are also hydrophobic and are one of the most widespread devices. In this case, the microstructuring has also a supplementary antireflective property, which improves the energetic yields during the conversion of energy.

4.3.2. *Optical properties*

The passage of light through two transparent materials, separated by a flat surface, gives rise to two separate phenomena: reflection and refraction.

Figure 4.8. *The self-cleaning principle of a microstructured surface. Due to the inclining surface, the water droplet rolls over the superhydrophobic surface carrying with it the specks of dust on the material (adapted from Hsu and Sigmund [HSU 10])*

The reflected beam of light is symmetrical to the incident beam, and the refracted beam (the beam that passes into the second milieu) is different to the incident beam predicted from an angle depending on respective refractive indices of the two domains. The angles of reflection and refraction are defined by a simple trigonometry relationship known as the Snell–Descartes law[8] or the law of refraction (Box 4.2).

When the material is structured in thin layers of distinct refractive indices, or even when it has a bumpy surface or regularly spaced holes with different indices, the passage of light is even more complicated. New properties are observed, which, depending on the type of structuring, lead to iridescence (a change in color of the surface according to the observation angle), to the complete reflection or to the absence of reflection, to the disappearance of certain wavelengths after crossing the structured medium or to the light being guided in specific directions.

8 Snell and Descartes are the two physicists who expressed the reflection and refraction laws of light. The British usually refer to Snell's laws, whereas Descartes' laws are preferred in France. It seems more appropriate to consider that both scientists discovered the laws simultaneously but independently.

Box 4.2. Snell–Descartes laws

Snell–Descartes laws, established in the 16th Century, describe the changes in direction of light rays crossing two transparent materials characterized by the refractive indices n_1 and n_2. The refractive index of a medium is defined by the relationship of light propagation speeds in a vacuum (c) and in the medium (V), that is $n_1 = c/V_1$, $n_2 = c/V_2$.

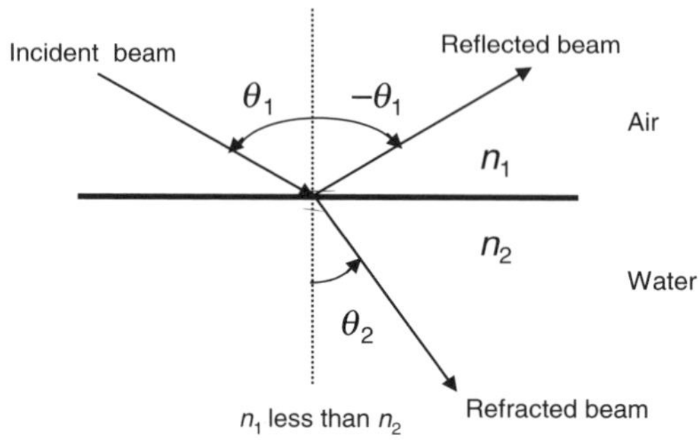

Snell–Descartes laws indicate that the reflected rays are symmetrical to the incident rays compared with the original ray at the interface. The refracted rays that penetrate the second medium are subjected to a deviation that verifies the trigonometric relationship $n_1 \sin \theta_1 = n_2 \sin \theta_2$.

In the opposite case where light passes through a medium (medium 1) of a higher index 2 (water) toward another medium (medium 2) with a lower index 1 (air), the refraction into air is suppressed as soon as the incidence angle increases beyond a particular value θ_0 ($n_2 \sin \theta_0 = n_1$). This is the total reflection phenomenon: the light becomes trapped in medium 2. It is precisely this principle that fiber optics are based on, used to guide light over long distances.

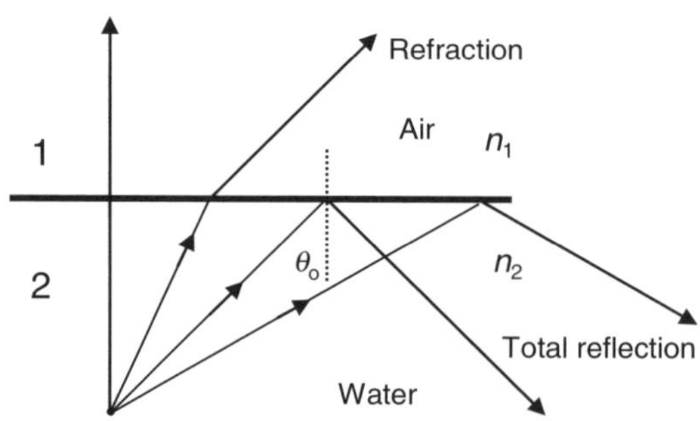

All these properties are widely exploited in the development of optoelectronic devices, such as wave guides, solar panels, antireflective surfaces and various devices similar to photonic crystals.

Industrial processes that have been used up until the present day to develop these materials essentially take these depositions (via evaporation of thin layers under vacuum conditions) techniques into account; they are particularly useful, for example, in the manufacture of antireflective glass.

More recently, numerous research groups have shown that it is possible to create a 3D nanostructure of materials by using colloidal lithography techniques. These techniques do not require heavy machinery and are therefore much less tedious than traditional evaporation techniques; they are well adapted to the manufacture of photonic crystals.

4.3.2.1. *Photonic crystals*

These are nanostructured materials with a periodic alternating of refractive indices.

Depending on the type of structure, there can be 1D, 2D or 3D photonic crystals (Figure 4.9).

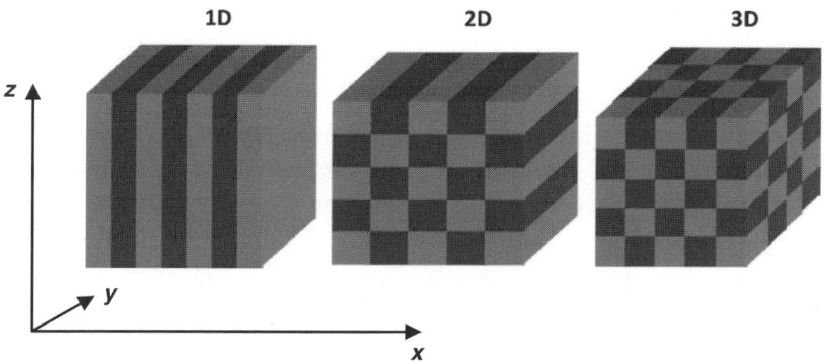

Figure 4.9. *Examples of one-dimensional (1D), two-dimensional (2D) and three-dimensional (3D) photonic crystals. The two colors (black and gray) correspond to two materials with different refractive indices. The 1D crystals are produced from alternating depositions of the two materials into thin layers, 1D indicating one periodicity according to direction x; the 2D crystals (periodicity in the xz plane) are formed by piling up layers of both materials in the xy plane. The 3D crystals correspond to nano- or micro-islands of both materials regularly distributed in space, the periodicity being established according to the three directions*

Although they have been recognized for a long time, they are still the subject of numerous studies and developments for important practical applications. Some applications include ultrahigh reflectivity mirrors that are particularly useful for improving the light emitted from lasers; the construction of low-loss waveguides that allow light to be channeled into optoelectronic devices used in telecommunications; the capturing of light by solar cells; and recently the design of tiny highly sensitive sensors to detect gases (Lourtioz *et al.* [LOU 05]).

Creating 1D and 2D photonic crystals (periodicity according to one or two directions) is not a new concept; however, 3D networks (periodicity in space according to three main directions) are much more recent.

It is the colloidal lithography techniques that have greatly contributed to their development, particularly by making a 3D structure of materials possible, with dimensional periodicities between 400 and 1,000 nm, compatible with visible light.

The main property of these photonic crystals is to selectively transmit electromagnetic radiation when the materials forming them are periodically distributed in space. Antireflective surfaces and waveguides are the main resulting applications.

4.3.2.2. *Waveguides*

Fiber optics, whose industrialization began in the 1970s, is currently one of the most successful methods of long-distance data transmission using optics. The principle is simple and is based on the fact that light propagates in the glass fiber under conditions of total reflection (Box 4.2). It therefore moves about by zigzagging against the glass walls over considerable distances, without a significant decrease in intensity, and without the propagation being affected by the curvature of the fibers, provided that this is not too excessive[9].

A completely different mechanism is used with photonic crystals to guide the light. In fact, their main feature is to act as a filter of electromagnetic radiation, by preventing a range of wavelengths from propagating inside the crystal, according to the geometric characteristics of the crystal and the

9 Fiber optics was first used in the 1950s with the first fibroscope being used in endoscopy examinations. It was not until after the Laser was invented in the beginning of the 1960s that the first telecommunication experiments began; they were developed extensively when manufacturers began to produce quality glass fibers in large amounts, a technological advancement made by the National Center of Telecommunications (CNET) in Lannion in the 1970s.

indices that make up the material. It is due to this particular property that they are often compared to "photonic semiconductors"; the forbidden band of wavelengths is similar to the forbidden energy band for electrons in SCs[10].

How to create 1D photonic crystals with alternating layers of different indices has been known for a very long time. This manufacturing technique usually uses the conventional evaporation–deposition procedure (of different materials under vacuum conditions), which permits the thickness and the flatness of successive layers to be accurately controlled. Better known as Bragg mirrors, these photonic crystals are perfect reflectors of a specific wavelength.

They are therefore frequently used as the coating of optical fibers to increase their reflectivity in the infrared domain and to decrease loss of light by as much as possible, inevitable over long propagation distances (optical fibers in telecommunications use infrared radiation with a wavelength of 1.5 µm). They are also involved in the construction of lasers to improve the reflectivity of the cavity walls, on which the beam of light is reflected before being directed to the outside.

Unlike 1D photonic crystals that have been greatly studied, developing 2D and 3D crystals, which are capable of interfering with visible light between 400 and 800 nm, is much less straightforward. Naturally occurring opals are formed from the assembly of microbeads, and can be compared to 3D photonic crystals; they have numerous defects in their crystals that contribute to their varied pearlescent colors, hence making them gemstones.

These 2D and 3D crystals have rather attractive properties for diverse future applications. The function of a perfect waveguide is certainly predicted to have a large number of prospective future applications, among the numerous lines of research at present, as well as in the domain of sensors that integrates optoelectronics.

The idea is to create a defect in a photonic crystal, such as deleting a row of nano-pads, which causes the light of a specific wavelength to be channeled through the crystal, following this pathway only (Figure 4.10).

---

10 It was Lord Rayleigh who observed the first forbidden band of light in 1887 in a 1D photonic crystal.

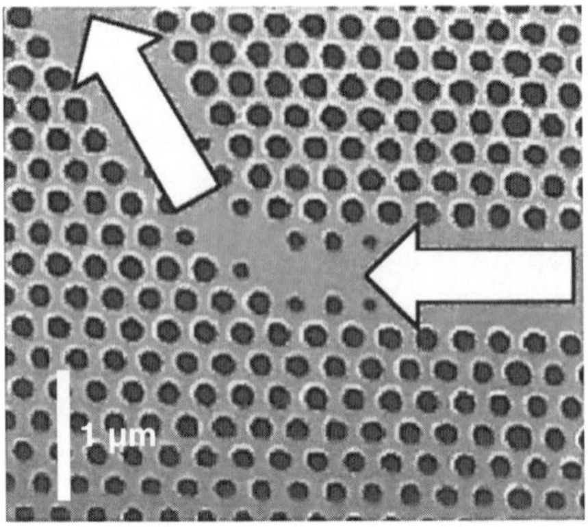

Figure 4.10. *Pathway of light in a two-dimensional photonic crystal acting as a waveguide (adapted from Lourtioz et al. [LOU 05])*

Light can also be guided almost without any decrease in intensity through pathways with 90° bends, which does not occur with a traditional optical fiber.

Creating waveguides of light in 3D photonic crystals is much more delicate. Considering that colloidal lithography techniques are being used, it is likely that practical implementations will soon appear.

4.3.2.3. *QDot LASER diodes*

The Light Amplification by Stimulated Emission of Radiation (LASER) effect (Box 4.3) was demonstrated with a ruby crystal for the first time in 1960 by the American physicist T. Maiman. Shortly after, Alferov and Kroemer[11] showed that a LASER beam could also be produced by applying a potential difference across an electronic compound composed of two linked doped SCs n and p[12].

---

11 Z.I. Alferov (Russian) and H. Kroemer (German, but recognized in the United States) shared the Nobel Prize of Physics in 2000 with J.S. Kilby (integrated circuits) for their work on heterostructured transistors.

12 A doped SC is different from an intrinsic SC (equal number of + and – charge carriers) in the fact that chemical "impurities" have been introduced in the crystalline network of the SC, which causes an excess of negative (SC n) or positive (SC p) charges to be produced. Al_XGa_{1-X} As ($0 < x < 1$) is a wide forbidden band SC unlike gallium arsenide (GaAs) and can be n or p doped depending on its composition.

Box 4.3. The LASER effect

The LASER effect is a "stimulated" emission of photons leading to a monochromatic and coherent beam of light (photons emitted in a synchronized fashion) with a very high intensity, due to the amplification phenomenon.

An emission of photons usually corresponds to a radiative transition of an electron between two energy levels E_1 and E_0 or to an electron-hole combination. This phenomenon can be observed with an atom, a molecule, a crystal or a semiconductor (SC). Unlike the general phenomenon of luminescence (Box 3.1) where the emission of light is said to be spontaneous and incoherent (photons randomly emitted), LASER emission originates from a so-called "stimulated" coherent emission of light, phenomenon that was described by Einstein in 1917.

Stimulated emission is produced when an electron in the excited state E_1 is subjected to the action of an incident photon of energy $h\nu$ ($h\nu = E_1 - E_0$). It returns to its fundamental state E_0 by emitting two photons $h\nu$, which have the same frequencies as the incident photon and the same phase-shift (synchronized waves). The amplification is due to the fact that there is only one incident photon but two photons are emitted.

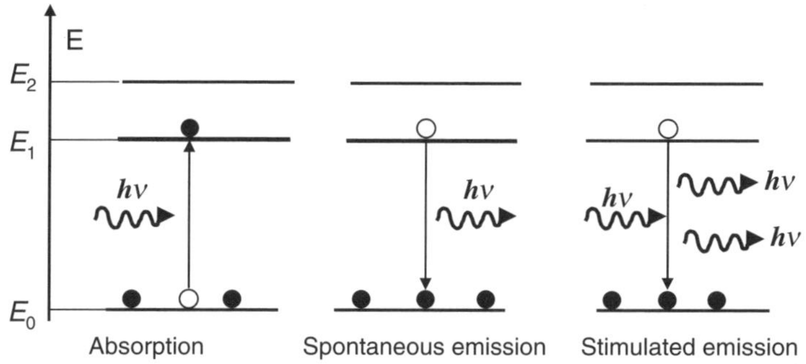

Stimulated emission, however, can only be observed if there is a "population inversion", that is a much higher number of electrons in the excited state E_1 compared to E_0 (the opposite situation to normal). This population inversion is carried out by a "pumping" process that can be either optical or electrical (in the case of an SC). It involves exciting electrons up to level E_2, much higher than E_1, and storing them at level E_1 that is more stable than E_2, from which stimulated emission is induced by an incident photon $h\nu$.

Significant progress was made at the end of the 1960s with systems based on a double heterojunction of two "wide band gap" (wide forbidden band) SC n and p separated by a non-doped "narrow gap" SC containing the active layer.

Subsequently, between 1975 and 1990, this structure developed into LASER diodes in thin layers, then moving toward "quantum well" LASER diodes; the thickness of the "narrow band" SC, sandwiched between the two wide band gap SC n-AlGaAs and p-AlGaAs, is approximately 10 nm.

The LASER effect is produced by combining the electrons and holes produced and accumulated in the narrow gap SC (active layer). From a current intensity threshold value, an accumulation of positive and negative charges is produced, leading to a population inversion from which the stimulated emission can be produced, giving rise to LASER effects. The photons emitted are also amplified by being successively reflected in the active layer, which acts as a cavity resonator matching the wavelength of the light emitted (Figure 4.11).

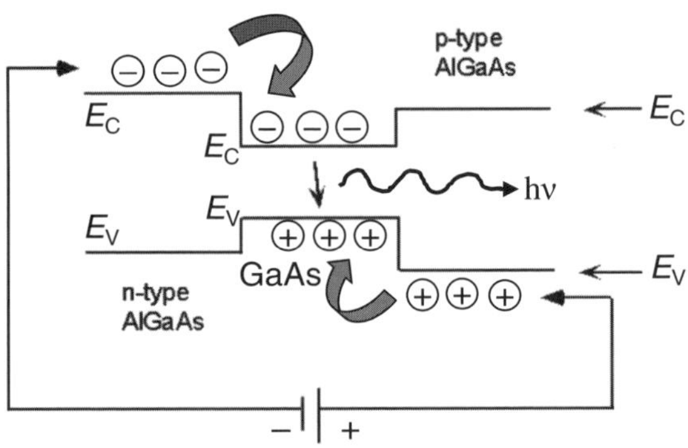

Figure 4.11. *Simplified energetic diagram of a double heterojunction LASER diode (n-AlGaAs/GaAs and GaAs/p-AlGaAs): n and p are large band-gap SCs (forbidden band) whereas the active emission region is a narrow band-gap SC (GaAs). E_C and E_V are the respective lower and higher energy levels of conduction and valence bands of different SCs. Beyond a current intensity threshold, the direct polarization causes electrons (– charges) and holes (+ charges) to accumulate, respectively, in the conduction and valence bands of the GaAs SC. This creates a "population inversion" (excess of negative charges and holes in the narrow band-gap SC). Combining electrons and holes then leads to what is called a "stimulated" emission corresponding to the LASER effect, with the emission of monochromatic radiation hν (hν = $E_C - E_V$) (Adapted from Diwu et al., [DIW 07])*

This device is extremely compact (with a size of approximately 1 mm) that explains its common place use in multiple domains (industry, medicine, etc.). In 2003, the LASER diode market was already worth approximately six billion dollars.

From a practical point of view, a heterojunction LASER diode is formed by successively stacking onto a GaAs substrate, a layer of AlGaAs (n), an active layer of non-doped GaAs and then a layer of AlGaAs (p).

A direct polarization is applied (+ and – terminals of a current generator connected to layers p and n, respectively). The active layer usually has a higher refractive index than SCs p and n, which makes it behave like a waveguide, which channels the beam of light into the active layer, reflecting it many times, before it is finally emitted to the exterior (Figure 4.12).

However, there is a problem regarding the consumption of energy, which is still very high. This translates as current density thresholds of approximately 50–80 A/cm^2, that is for a device such as that shown in Figure 4.12 with intensities of several hundred milliamperes (Bouchoule [BOU 10]).

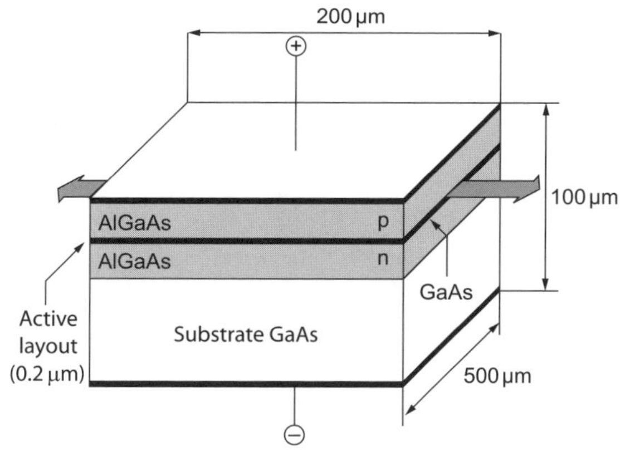

Figure 4.12. *Diagrammatic representation of a double heterostructured LASER diode. Note the tiny size of the device, less than 1 mm, and the active layer has a thickness of approximately 200 nm. With active layers of several nanometers in thickness, the device is considered as a quantum-well LASER diode (Adapted from Prasad, [PRA 04])*

Over the past decade, new research pathways have been developed to reduce the intensity of current required to produce a LASER effect. Replacing quantum wells by self-assembled QDots, which is expected to significantly reduce the current intensity trigger threshold of the LASER effect, has already been the subject of numerous experiments at the laboratory scale. Recent results have verified this improvement as current intensity thresholds less than 10 A/cm^2 were obtained, which involve an important improvement with regard to older devices, with threshold current intensities of around several hundred amperes per square centimeter (Bouchoule [BOU 10]).

The manufacturing process of self-assembled QDots is rather tricky and is currently based on the use of relatively complex vapor deposition techniques in vacuum conditions, which require further simplification to be able to proceed in putting this device on the market.

Chemical vapor deposition (CVD) techniques used in the SC industry represent a first improvement. They allow the deposition of thin layers onto a solid substrate, by causing vaporized reactants to react together in a partial vacuum, by clumping together. QDot networks with good geometric characteristics are obtained according to this procedure.

Figure 4.13 represents an AFM image of a layer of InAs QDots (Indium arsenide, narrowband SC) deposited onto a GaAs substrate and obtained by metalorganic chemical vapor deposition (MOCVD) that involves using arsenic and indium metalorganic (MO) compounds as reactants.

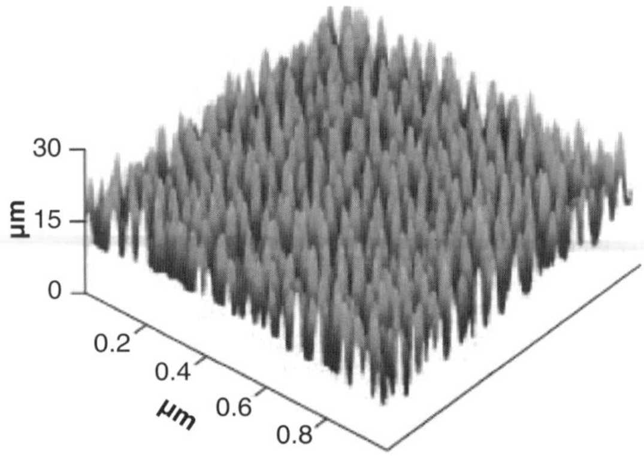

Figure 4.13. *AFM image of a layer of InAs QDots (narrow band-gap) on a GaAs substrate obtained using the MOCVD technique (the number of QDots is 5.10^{10} /cm^2 (i.e. 50 billion QDots per cm^2). These QDots are pyramid shaped, with an average height of 10 nm and a diameter at the base of approximately 25 nm (adapted from Guimard et al. [GUI 06])*

These results once again show how interesting it is at present to organize NPs into regular networks on solid surfaces, in particular, by using colloidal lithography. Applying this to the construction of LASER diodes is still under development, but we can imagine that new production techniques of networks like these will appear in the near future and should provide innovative solutions to the large-scale manufacturing of these devices.

4.3.2.4. *Antireflective surfaces*

Antireflection is an extremely important property of surfaces. It allows, along with other applications, a considerable improvement in the amount of light collected on photovoltaic cells, and ultimately their yield of converted energy (see Chapter 6).

It is estimated that Si surfaces, with very high refractive indices, reflect approximately 30% of the incident light. Usually, this is overcome by depositing (in a vacuum) various thin layers with an increasing refractive index, calculated to ultimately reduce the reflection.

One way of reaching this objective is to draw inspiration from natural examples seen in a large number of nocturnal insects.

Nature often creates beautiful things. The eyes of certain nocturnal butterflies are equipped with an antireflective cornea. This is not composed of a multilayered structure, but instead appears to be made up of tiny bumps of approximately 200 nm, arranged in a regular fashion over the surface, the spitting image of an antireflective 2D photonic crystal (Figure 4.14).

Figure 4.14. *Image showing the cornea of a nocturnal butterfly native to East Asia (Polygonia c-aureum). The periodic variation of refractive index between trapped air and the bumps cause a significant reduction in the intensity of the light reflected (adapted from Stavenga et al. [STA 06])*

This type of structure can be easily created using colloidal lithography techniques, much less tedious than those used in the microelectronics industry, and are just as efficient (see Chapter 3, colloidal lithography, Figure 3.18).

Applying this principle to silicon surfaces, nanostructured by a colloidal monolayer, is rather significant and shows much better antireflective properties than silicon treated according to conventional techniques (Figure 4.15).

These nanostructuring techniques are still under development, but given the constant progress being made in developing this process; large-scale use will not be far off; it could also contribute financially in the market of photovoltaic cells.

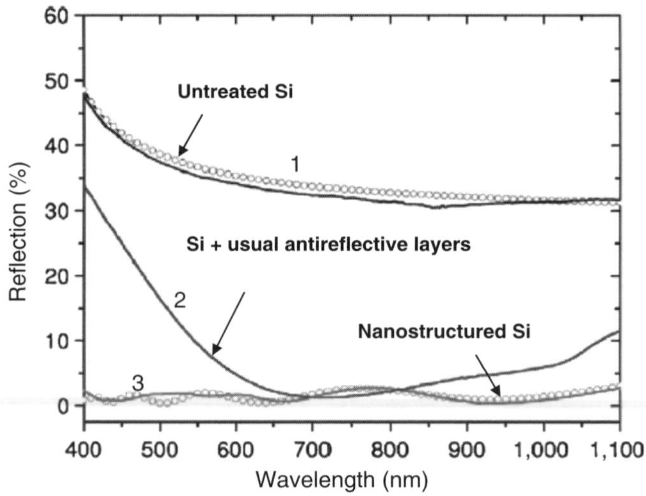

Figure 4.15. *Comparison of antireflective properties of silicon surfaces – 1: untreated surface; 2: Si surface of a commercial photovoltaic cell conventionally treated with a deposition of an antireflective coating based on silicon nitride (SiN); and 3: silicon surface nanostructured using a colloidal lithography technique. Note, in the last example, a very definite improvement in the antireflective properties throughout the whole visible spectrum between 400 and 600 nm (adapted from Sun et al. [SUN 08])*

4.3.2.5. *Plasmonic crystals and the SERS effect*

Plasmonic crystals are 2D networks of metallic nanostructures, usually formed from a regular assembly of NPs or nanoholes; in the case of metals such as gold or silver, they are known to have an intense PR when excited by visible light. The formation of a network leads to localized PRs, which can be controlled by adjusting the size, the spacing and the shape of the NPs.

Practically, these are mainly used as sensors to detect and titrate particular molecules in tiny amounts.

The phenomenon observed is the displacement of the plasmonic resonance (PR) by only a few nanometers, caused by a modification in the immediate environment of the metallic surface, for example by the adsorption of a molecule. This can be exploited to detect traces of molecules, and in this case the concentration of the chemical compound in solution can be calculated according to the displacement of the wavelength of the PR.

This is exactly what is observed with the avidin–biotin titration[13], two molecules that associate, due to biotin's affinity for avidin, and together they form a highly stable supramolecule.

In order to carry out this titration using the PR method, a structured silver surface is used (obtained with colloidal lithography) on which biotin molecules bind until saturation (Figure 4.16). The addition of a solution of avidin leads to the formation of a biotin–avidin complex, and ultimately to the shift of the PR. By measuring the variation in the wavelength corresponding to the PR, a linear increase in this wavelength is observed according to the concentration of avidin added. The detection sensitivity is extremely good; it is approximately 100 and 400 picomoles (100 picomole or 0.1 nanomole equivalent to 60 billion molecules per ml).

We can also collect more specific information using SERS, directly linked to PR and resulting from a strong increase of the electrical field localized in the environment right next to the surface. Scattered light, partially absorbed by molecules, gives a spectral "fingerprint" of each molecule; this allows it to be identified, and at the same time its concentration to be determined. This technique is being used more and more to detect biological molecules present in very low concentrations.

Thus, in the case of a particular component such as four-aminothiophenol ($H_2N–C_6H_4–SH$ or 4–ATP), adsorbed onto silver, by using the same type of

---

13 Biotin is a small molecule known as vitamin B8 (also known as BH) that forms an association complex (supramolecule) with avidin (a protein), which is much larger than biotin.

nanostructured silver surface, the authors observe a signal amplification about 10 million times greater than with simple Raman scattering (Hong *et al.* [HON 10]).

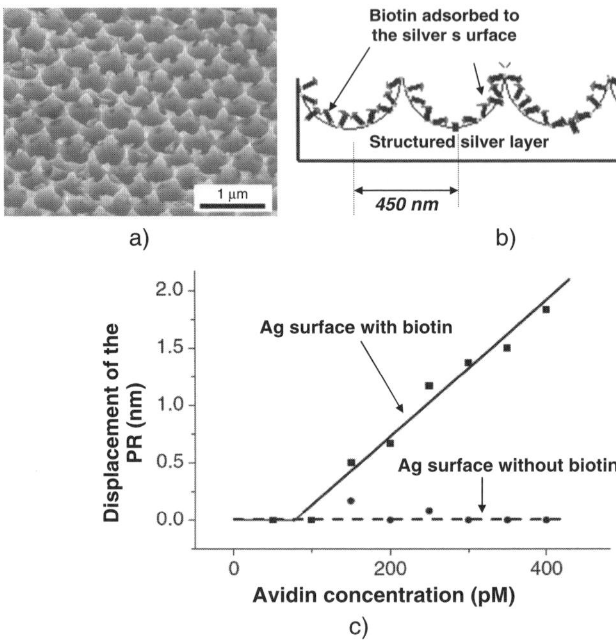

Figure 4.16. *Avidin titration by the shift of the PR of a silver surface structured and modified by biotin: a) TEM image of a nanostructured Ag film formed using colloidal nanolithography; b) diagrammatic representation of a silver surface structured and modified by biotin; c) variation in the displacement of the wavelength of the PR according to the concentration of avidin added: when no biotin is present the PR does not change, when biotin is present, avidin binding to biotin causes a linear displacement in the PR (adapted from Hong et al. [HON 10])*

Figure 4.17 represents three Raman scattered spectra with 4-ATP corresponding to the Raman scattered spectrum of 4-ATP in the solid state, and to two SERS spectra corresponding to a molecular layer of 4-ATP attached through the sulfur atom onto the nanostructured silver surface. The different peaks measured in cm^{-1} ($1/\lambda$) correspond to the scattered light and reflect different vibrating patterns of molecules and their positions; this is like a molecular fingerprint, and therefore permits its identification.

This SERS technique is probably one of the most sensitive detection methods. It has developed considerably regarding its performance. The detection sensitivity keeps on improving, and numerous researchers have already managed to obtain a SERS spectrum for a single molecule; this could revolutionize techniques for analyzing compounds in trace amounts in the future (Camden *et al.* [CAM 08]).

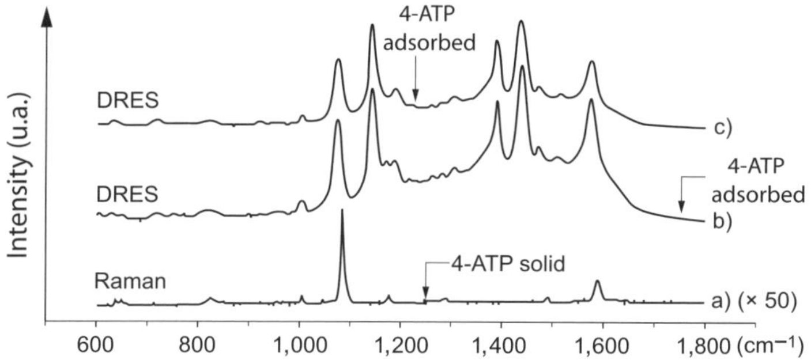

Figure 4.17. *Four-aminothiophenol (4-ATP) SERS and Raman spectra: a is the Raman spectrum that was obtained from 4-ATP in the solid state. It has a low intensity (magnification X50), whereas the SERS spectra obtained from a single molecular layer of 4-ATP fixed onto a nanostructured silver surface are considerably greater. The SERS spectra b and c do not differ considerably in intensity, even though the two nanostructured silver surfaces differ in spacing (b: 130 nm, c: 450 nm) (adapted from Hong et al. [HON 10])*

4.3.3. *Nanoelectronics applied to the detection of trace elements: nanowire transistors*

The introduction of CNTs, as well as new opportunities to develop semiconducting NWs of only a few nanometers in diameter (but with lengths reaching several microns, or even several millimeters), has provided numerous prospective applications toward the design of new transistor "chips" that can be used in the detection of trace molecules.

The first sensors using SCs made of silicon NWs, which are able to detect compounds in trace amounts, appeared right at the beginning of the year 2000; the inventors were C.M. Lieber and his team at Harvard University in the United States [CUI 01].

These devices are based on the FET in which the NW is the SC linked to two electrodes "Source" and "Drain" (see Box 5.1, MOSFET). These nanotransistors can be arranged into a network, which allows multiplexed analyses to be carried out, by simultaneously measuring the variations in conductivity caused by the adsorption of diverse chemical or biological molecules to the surface of the NWs. The simplicity of this technique, the high detection sensitivity along with great specificity, makes them particularly attractive as sensors when it comes to carrying out control operations on land in the absence of any heavy logistics, especially in crime or environmental investigations.

4.3.3.1. *The operating principle of the FET sensor*

The majority of molecules that adsorb onto a surface acquire electric charges and these charges are localized at distances of only a few nanometers from the surface of the SC. This proximity of charges with regard to the surface of the SC induces an electric field in the NW, similar to that exerted by a FET electrode gate up to a potential of several volts but at much larger distances from the SC (several nanometers). So, depending on the molecules adsorbed onto the surface of the NW, significant variations in the source-drain current can be observed, therefore revealing whether certain molecules are present or not.

4.3.3.2. *An example of how it could be applied: detecting explosives*

The increase in terrorism worldwide has encouraged research into efficient devices for detecting the presence of a solid explosive by analyzing the air. This principle is based on the fact that most solids give off a vapor that mixes with air at very low concentrations (latent vapor pressure of solid). When it gives off an odor, the sense of smell must be highly developed for it to be detected, such as that of police dogs.

There are numerous rather complicated methods that allow the direct analysis of mixtures of gases. What is difficult is that the majority of solid explosives have very low vapor pressures that require heavy machinery capable of detecting tiny amounts of gas mixed in the surrounding air.

Trinitrotoluene (TNT), one of the most frequent explosives found in terrorist attacks, is characterized by a very low vapor pressure (7.6×10^{-9} Pa, that is 7.6 billionths of atmospheric pressure) at ambient temperature (25°C). The unit used by specialists is ppb (parts per billion); this corresponds to a concentration to air ratio of approximately 7 ppb. This value is very low and measuring it requires an installation such as the mass spectrometer[14], which is difficult to transport and is therefore usually set up in fixed positions.

A team of researchers at the University of Tel-Aviv in Israel (Engel *et al.* [ENG 10]) showed that this measurement could be done in a much simpler way, by using silicon NW networks, with aminosiloxanes ($H_2N(CH_2)_3Si(OC_2H_5)_3$ or APTES) bound to the surface.

These aminosiloxanes specifically bind to TNT molecules with electrostatic forces, leading to positive charges appearing on the surface of the N atom of NH_2, and negative charges on the benzene ring (Figure 4.18). It is these positive charges, closest to the silicon surface (a fraction of a nanometer from the surface), which will induce the electric field responsible for the variations in conductivity of the NW.

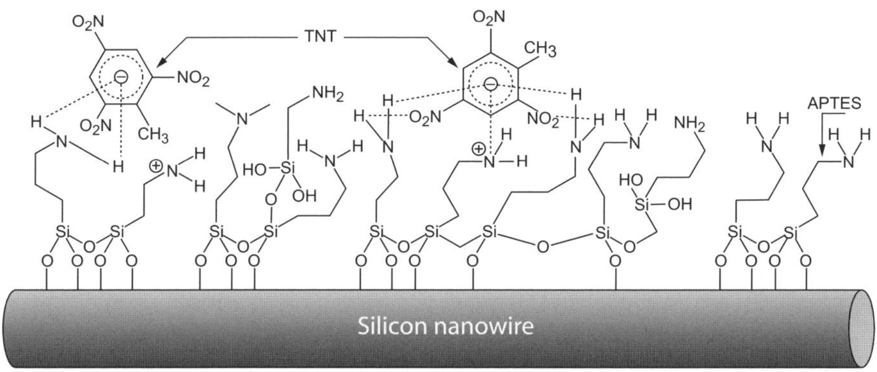

Figure 4.18. *Diagrammatic representation of APTES molecules binding to a p-Si nanowire. The interaction of the amino group (NH_2) with TNT leads to a highly charged complex with a negative charge localized on the ring and a positive charge delocalized in the surrounding NH_2 groups. The positive charges closest to the silicon surface cause the conductance of the nanowire to decrease (adapted from Engel et al. [ENG 10])*

This nanosensor can operate in the liquid state (the TNT is dissolved in a specific solvent) or directly in the gaseous state (the TNT mixed with air is directly put into contact with the NW).

14 Mass spectrometry is an analysis technique used in physics that allows a chemical compound to be identified by measuring its mass. Once the compound is introduced into the device as a vapour, it is fragmented into positive ions. Applying an electrical and a magnetic field allows the fragments to be separated and the molecules to be identified.

In this last case, on allowing short bursts of gas to enter the system, we observe each time gas enters a finite change in conductance[15], which continues until APTES is completely chemically saturated by TNT (Figure 4.19). However, note that, after two gaseous injections, the NW is unable to recover its initial conductance value, meaning that the adsorption of TNT is irreversible.

The sensitivity of the detector is remarkable: a relative change of 10% in the conductance is registered from an injection of only 5 ppt (5 parts per trillion, that is a concentration ratio between TNT and air of 1 in 200 billion); this represents a concentration of TNT 1,000 lower than that corresponding to the vapor pressure at ambient temperature (5 ppb).

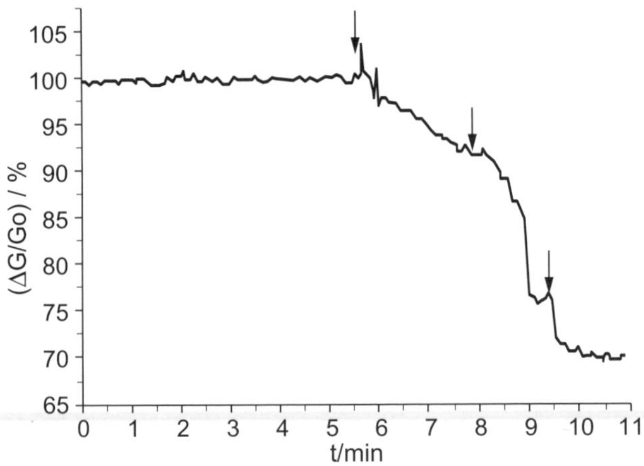

Figure 4.19. *Response of a Si nanowire in real time (p doped, and modified by APTES) on adding successive bursts of a gaseous mixture of air + TNT (5 ppt) for 5 s. The arrows indicate when the gas is introduced into the nanosensor (adapted from Engel et al. [ENG 10])*

It must be noted that this detection limit of 5 ppt is attainable with only a few techniques, among which is SPR applied to an immunology reaction.

Using SC NWs for analytical purposes is quite recent, and the examples we have described are only a glimpse of their full capabilities. Incorporating them into field-effect transistor assembly gives a detection sensitivity similar

15 The conductance of a material (inverse of the resistance) is proportional to the current intensity flowing between the drain and the source.

to, if not better than, what is observed with the best analysis systems. Moreover, being easily employed in networks opens up new areas of possible exploitation, such as the design of "artificial noses" (also known as electronic noses) widely used in the agri-food sector (automatic detection of flavors and smells, food control, surveillance of production lines, detecting bacteria, etc.).

4.3.3.3. *Electronic noses*

"Electronic noses" are in fact systems of sensor networks that automatically detect an odor or a particular gas found in a gaseous mixture (Röck *et al.* [RÖC 08]). This measurement is based on recording a "fingerprint" corresponding to the assembly of different signals received by each sensor. The compound is identified with the help of a piece of detection software by comparing the obtained signal to those of known molecules, previously established and stored in a database as a reference. This operational mode is very similar to our olfactory system, which is composed of a large number of receptor cells, sensitive to certain gaseous odors, subsequently identified by our brain.

Artificial noses use a large number of conventional SC sensors based on metallic oxides, which are not very specific. However, SC NWs with surfaces chemically modified by specific molecular receptors are extremely sensitive, and can produce electrical signals that can be easily measured. Creating NW networks using traditional lithography methods in vacuum conditions, along with the many possible modifications that can be carried out on the surface, opens up new prospective pathways in the analysis and detection of compounds in tiny quantities. In the future, they could be used in the construction of performing artificial noses, with flexible uses and low costs.

4.4. Conclusions and perspectives

Organizing nanomaterials into regular networks is a source of new properties, very different to individual nanomaterials. Very diverse fabrication techniques can be adapted to the geometric characteristics and dimensions of the networks that we wish to construct.

At the nanometric scale, the self-assembly techniques of molecules, controlled by the formation of hydrogen bonds, generate extremely homogeneous cavities on a surface, in which particular reactants or nanomaterials can be inserted.

For intermediate sizes of several nanometers, using porous membranes as a mask is useful for the development of nanomaterials in diverse shapes, in the same way that networks can be formed by applying self-association techniques with bifunctional chemical mediators capable of bonding with their ends to metallic or SC NPs.

For sizes greater than 100 nm, recently developed colloidal lithography has undoubtedly proved to be promising in the production of highly regular 2D and 3D networks of particles.

All of these networks of nanomaterials have properties that differ considerably compared with those of isolated nanomaterials. Among the most important, PR, the SERS effect, antireflective properties, light wave guides, superhydrophobic or superhydrophilic properties of surfaces, are the properties that are enhanced by the regular structuring of surfaces and the formation of 2D or 3D networks of particles.

Significant progress has been made recently by simplifying the techniques used to create 2D and 3D photonic crystals. Developing colloidal lithography techniques could provide easy access to the structuring of surfaces and to the creation of 2D and 3D micronic and submicronic networks. Among the numerous possible applications, forming QDots networks to produce LASER diodes could be an advantageous alternative compared to heavy microelectronics techniques.

4.5. Bibliography

[BOU 10] BOUCHOULE S., *Les nouvelles diodes laser pour les télécoms optiques*, Doc CNRS, 2010.

[CAM 08] CAMDEN J.A., DIERINGER J.A., ZHAO J., VAN DUYNE R.P., "Controlled plasmonic nanostructures for surface-enhanced spectroscopy and sensing", *Accounts Chemical Research*, vol. 41, pp. 1653–1661, 2008.

[CHE 08] CHEN M., GOODMAN D.W., "Catalytically active gold on ordered titania supports", *Chemical Society Reviews*, vol. 37, pp. 1860–1870, 2008.

[CUI 01] CUI Y., WEI Q., PARK H.K., LIEBER C.M., "Nanowire nanosensors for highly sensitive and selective detection of biological and chemical species", *Science*, vol. 293, pp. 1289–1292, 2001.

[DAN 04] DANIEL M.-C., ASTRUC D., "Gold nanoparticles: assembly, supramolecular chemistry, quantum-size-related properties, and applications toward biology, catalysis, and nanotechnology", *Chemical Reviews*, vol. 104, pp. 293–346, 2004.

[DIW 07] DIWU H., ARDA B., ECE 580, Term project, quantum dot lasers, 2007, available at http://www.ece.rochester.edu/courses/ECE580/docs/Quantum_ Dot_ Lasers.pdf

[ELG 97] ELGHANIAN R., STORHOFF J.J., MUCIC R.C., LETSINGER R.L., MIRKIN C.A., "Selective colorimetric detection of polynucleotides based on the distance-dependent optical properties of gold nanoparticles", *Science*, vol. 277, pp. 1078–1081, 1997.

[ENG 10] ENGEL Y., ELNATHAN R., PEVZNER A., DAVIDI G., FLAXER E., PATOLSKY F., "Supersensitive detection of explosives by silicon nanowire arrays", *Angewandte Chemie, International Edition*, vol. 49, pp. 6830–6835, 2010.

[GUI 06] GUIMARD D., TATEBAYASHI J., ARAKAWA Y., "Diodes laser pour les télécommunications optiques", *Techniques de l'Ingénieur*, vol. 8, no. RE 48, pp. 1–12, 2006.

[HAY 01] HAYNES C.L., VAN DUYNE R.P., "Nanosphere lithography: a versatile nanofabrication tool for studies of size-dependent nanoparticle optics", *Journal of Physical Chemistry B*, vol. 105, pp. 5599–5611, 2001.

[HON 10] HONG G., LI C., QI L., "Facile fabrication of two-dimensionally ordered macroporous silver thin films and their application in molecular sensing", *Advanced Functional Materials*, vol. 20, pp. 3774–3783, 2010.

[HSU 10] HSU S.-H., SIGMUND W.M., "Artificial hairy surfaces with a nearly perfect hydrophobic response", *Langmuir*, vol. 26, pp. 1504–1506, 2010.

[JIA 09] JIA L., CAI W., WANG H., SUN F., LI Y., "Hetero-apertured micro/nanostructured ordered porous array: layer-by-layered construction and structure-induced sensing parameter controllability", *ACS Nano*, vol. 3, pp. 2697–2705, 2009.

[LE 04] LE J.D., PINTO Y., SEEMAN N.C., MUSIER-FORSYTH K., TATON T.A., KIEHL R.A., "DNA-templated self-assembly of metallic nanocomponent arrays on a surface", *Nano Letters*, vol. 4, pp. 2343–2347, 2004.

[LEI 11] LEI Y., YANG S., WU M., WILDE G., "Surface patterning using templates: concept, properties and device applications", *Chemical Society Reviews*, vol. 40, pp. 1247–1258, 2011.

[LOU 05] LOURTIOZ J.-M., BENISTY H., BERGER V., GERARD J.-M., MAYSTRE D., TCHELNOKOV A., *Photonic Crystals*, Springer, Berlin, Heidelberg, New York, 2005.

[PRA 04] PRASAD P.N., *Nanophotonics*, J. Wiley & Sons, 2004, available at http://www.ece.rochester.edu/courses/ECE580/docs/Quantum_Dot_Lasers.pdf

[RÖC 08] RÖCK F., BARSAN N., WEIMAR U., "Electronic nose: current status and future trends, *Chemical Reviews*, vol. 108, pp. 705–725, 2008.

[ROG 11] ROGACH A.L., "Fluorescence energy transfer in hybrid structures of semiconductor nanocrystals", *Nano Today*, vol. 6, pp. 355–365, 2011.

[ROS 05] ROSI N.L., MIRKIN C.A., "Nanostructures in biodiagnostics", *Chemical Reviews*, vol. 105, pp. 1547–1562, 2005.

[STA 06] STAVENGA D.G., FOLETTI S., PALASANTZAS G., ARIKAWA K., "Light on the moth eye corneal nipple array of butterflies", *Proceedings of the Royal Society B*, vol. 273, pp. 661–667, 2006.

[SUN 08] SUN C.-H., LIANG P., LIANG B., "Broadband moth-eye antireflection coatings on silicon", *Applied Physics Letters*, vol. 92, 061112, 2008.

PART 2

Applications and Societal Implications of Nanotechnology

It has long been an axiom of mine that the little things are infinitely the most important.

Sir Arthur Conan Doyle

Chapter 5

Nanoelectronics of the 21st Century

5.1. Some history

The original transistor, invented in 1947 by Shockley, Brattain and Bardeen (Nobel Prize, awarded 1957) from the Bell Laboratories in the United States, was by no means a miniaturized electronic device, yet this invention resulted in a radical technological revolution. Seemingly unable to threaten the vacuum tube industry (lamps, triodes, tetrodes and pentodes), many scientists were skeptical about the usefulness of replacing the old radio vacuum tubes with such devices.

It took more than 20 years before the use of transistors became widespread, when they started being produced in the form of the thin film electronic chips that we use today.

This required considerable progress in various fundamental areas of research, particularly in the production of thin film materials; automated production of thin film materials on a large scale resulted in the appearance of a new branch in the electronics industry. The Intel® Corporation, founded in 1968 by a small start-up team, was the first in the United States to present in 1971 a microprocessor consisting of several thousand transistors[1] (Figure 5.1).

---

1 Intel® was founded in 1968 by Gordon Moore and Robert Noyce, both of whom were previously engineers at the American company Fairchild Semiconductor.

Figure 5.1. *The first microprocessor (Intel® 4004-86), consisting of 2,300 transistors, made in 1971. The engravings have a thickness of 10 μm*

Building upon the original invention of Shockley, Brattain and Bardeen, other technological revolutions followed. First came the invention in 1958 of the integrated circuit by Texas Instruments' Jack Kilby (for which he won the Nobel Prize in 2000), in which all the components surrounding the transistor are made from the silicon substrate. This was followed in 1961 by the creation of field-effect transistors (FETs), which considerably simplified further transistor development. This eventually led to the CMOS, which is still used in the processors of modern computers; the base unit of the CMOS is the metal oxide semiconductor field-effect transistor (MOSFET) (Box 5.1 MOSFET).

It would be safe to say that the 1960s witnessed the birth of microelectronics. However, one must note that all the ground-breaking advances that took place during this period were only made possible by simultaneous progress in the understanding of the physicochemistry of the surfaces and interfaces of thin film materials; progress itself only being possible thanks to the increasing quality of the observational methods available.

The first scanning electron microscopes (SEM) and electron spectroscopy for chemical analysis (ESCA)[2] devices were commercialized in the 1960s, and enabled the precise study of the chemical phenomena that occur at the interfaces of materials in contact with each other. Without a doubt, this fundamental research allowed vast improvements in the techniques used in

2 ESCA, also known as XPS (X-ray photoelectron spectroscopy), is a photoelectric technique used in a surface chemistry analysis that allows the identification of the elemental composition of a pure state material to depths of up to 10 nm. The surface is irradiated with X-rays, resulting in the emission of electrons. The number of these is measured, along with their kinetic energy, which is characteristic of the atoms constituting the analyzed material and of the bonds between them. This technique was developed by Kai Siegbahn (Nobel Prize awarded 1981) in the 1960s.

the construction of thin film materials, and was highly important in the development of the microelectronics industry.

Box 5.1. Field-effect transistors: the metal oxide semiconductor field-effect transistor (MOSFET)

The MOSFET was first created using thin film in the 1960s by the Bell Labs. It comprises four electrodes: the *source (S)*, the *drain (D)*, the *gate (G)* and the *substrate* (n-type or p-type silicon, that is semiconductor (SC) consisting mainly of electrons (– charge, SC n) or holes (+ charge, SC p)).

The gate *G* is separated from the substrate by an insulating layer of silicon oxide. The gate exerts an electric field on the electrons of the SC, which either pulls them toward the interface between the SC and the oxide layer (if *G* has a positive + potential), or pushes them away from the interface (if *G* is negative). If a voltage $\mathbf{V_{DS}}$ is applied between the electrodes *S* and *D*, the electrons (represented by dots), present in the interface in a very narrow "channel", migrate from *S* to *D* (in the case of an SC n) and create an electric current $\mathbf{I_{DS}}$ in the opposite direction.

This current, $\mathbf{I_{DS}}$, is modulated by the voltage at the gate, $\mathbf{V_{GS}}$. $\mathbf{I_{DS}}$ is strong when $\mathbf{V_{GS}}$ is positive, the contrary when $\mathbf{V_{GS}}$ is negative. The highest intensity of $\mathbf{I_{DS}}$ is in the order of a million (10^6) times greater than its lowest intensity. These variations can be coded using the numbers 1 and 0, which makes the MOSFET adapted to the conception of digital circuits.

$\mathbf{V_{GS}}$ **(–)** → $\mathbf{I_{DS}}$ **(negligible)** → **signal (0)**
$\mathbf{V_{GS}}$ **(+)** → $\mathbf{I_{DS}}$ **(high)** → **signal (1)**

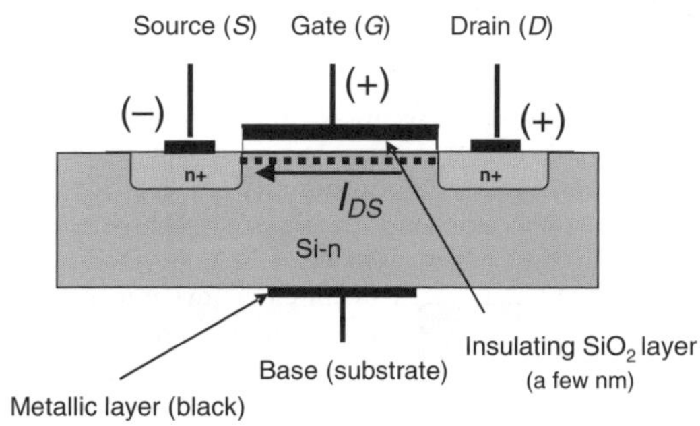

The areas marked n+ designate zones of silicon rich in electrons as opposed to the substrate, and are therefore highly conductive. The areas in black are metal deposits that provide a reliable electrical contact between the electrodes and the substrate. Combining two MOSFETs, one with an n-type substrate and one with a p-type substrate, results in a complementary metal oxide semiconductor (CMOS) device, which is more energetically advantageous.

Since then, these techniques have been progressing incessantly, such that today the lithography techniques used in the electronics industry are spectacularly miniaturized, due to the ever-decreasing width of the engravings used in the making of transistors. This has allowed the increase of the number of transistors over the same surface, responsible for the increased power of modern processors. Moore's Law, first evoked in 1965, predicts a doubling of the transistor density every two years; in other terms, this means a decrease in engraving width by a factor of 0.7 for each generation of transistor (Figure 5.2).

Remarkably, despite being first pronounced right at the start of the industrialization of microelectronics, Moore's Law has so far been accurate, and is a good description of the constant increase in performance of the industry until now.

Accordingly, the graph in Figure 5.2 predicts that in 2012, the width of the engravings should be roughly around 10 nm, with transistor dimensions of less than 100 nm. So far, this prognostic appears to be accurate.

Despite these impressive observations, Moore's Law cannot be true forever: the miniaturization of circuits has physical limits, both technological and theoretical. Below sizes of a few nanometers, the circuits will no longer be governed by the regular laws of physics. At this point, the concepts of quantum mechanics take over.

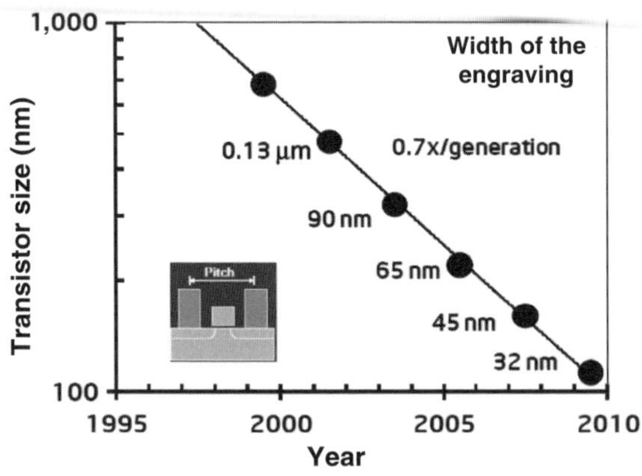

Figure 5.2. *Evolution of engraving width and transistor size made using the CMOS technique between 1995 and 2010 (adapted from an Intel® Corp. document)*

Other systems must therefore be considered. A scale of dimensions that would allow a decrease in size by a factor of 100 is the scale of the molecule, which leads to the domain of molecular electronics.

5.2. Molecular electronics

The 1970s saw the beginning of conceptual research in molecular electronics. Well aware of the inevitable physical limitations of silicon, several research teams began studying the electronic properties of molecules, with an underlying goal of creating molecular transistors.

There are two objectives behind this goal: first, to create new systems that will allow us to further increase the number of transistors per surface unit, and second, to decrease the energy used by computers using silicon-based components, present in all large servers.

The vast energy consumption of large servers is a major issue. The amount of digital information in the world is ever-increasing (a recent report by the International Data Corporation (IDC) indicates that between 2006 and 2011, the amount of digital information worldwide increased by a factor of 10, going from 180 to 1,800 EB[3]). Proportionally, energy consumption is amplified and becomes a serious problem. Recently, several organizations have looked into the matter, publishing guidelines concerning the optimization of IT servers (for example the 2011 report by the US Department of Energy, "Best practices guide for energy-efficient data center design", http://www1.eere.energy.gov/femp/pdfs/eedatacenterbestpractices.pdf).

This combination of considerations fully warrants research into systems that require less energy. Understandably, most new research in transistor and memory technologies have a strong focus on decreasing energy consumption and remains a top priority.

The major challenge is single-electron electronics, or simply "Single Electronics". At this extreme level of miniaturization, which is ultimately the molecular level, the regular laws of physics are no longer applicable and are replaced instead by the laws of quantum mechanics.

3 An exabyte (EB) is equal to a billion gigabytes, or 10^{18} bytes. One byte is itself made up of eight digits (bits), each digit corresponding to either a 1 or a 0. A byte can represent a letter, a number or a symbol. For the sake of comparison, current personal computers have a storage space of around 500 GB; http://france.emc.com/collateral/analyst-reports/diverse-exploding-digital-universe.pdf.

5.2.1. *Single Electronics. Dream or reality?*

The physicist Millikan[4] was the first to demonstrate, in 1911, that it was possible to manipulate individual electrons by measuring the electric charge of an electron that had been adsorbed by a droplet of oil in suspension in air.

Transferring or isolating small numbers of electrons within an electrical circuit, however, is far more complex. It was only in the 1980s that this became possible. Nowadays, it is possible to transfer electrons one-by-one into a small *conductive island* or quantum dot, the dimensions of which are in the order of a few nanometers.

This kind of research is still largely limited to the lab; the possibility of replacing silicon devices with this kind of technology on an industrial level remains a thing of the distant future. Nonetheless, these experimental studies do allow a greater understanding of electronic devices once the confinement of matter becomes considerable.

5.2.1.1. *Electron box and electron transfer by quantum tunneling*

An electron box can be obtained by simply inserting a nanoscale metal particle (the conductive island) into an isolating material in between two metallic plates serving as electrodes. These electrodes are then connected to a source of voltage. The superior electrode is placed very close to the metal particle, allowing the transfer of electrons by quantum tunneling to take place. The inferior electrode is placed far from the dot, stopping the flow of direct current in the circuit (Figure 5.3).

But how do the electrons pass from the superior electrode to the conductive metal particle?

The transfer of the first electron is relatively simple. The insulator between the superior electrode and the particle being very thin, a single electron can tunnel through it by quantum tunneling, in a reversible manner (from the top downward or from the bottom upward).

4 R.A. Millikan (American physicist, who won the Nobel Prize in physics in 1923) is famous for having accurately measured the electric charge of an electron. His relatively simple experiment involved using a microscope to observe a small droplet of oil, electrically charged by electrons, and to determine for which value of electrical field intensity the particle became immobilized.

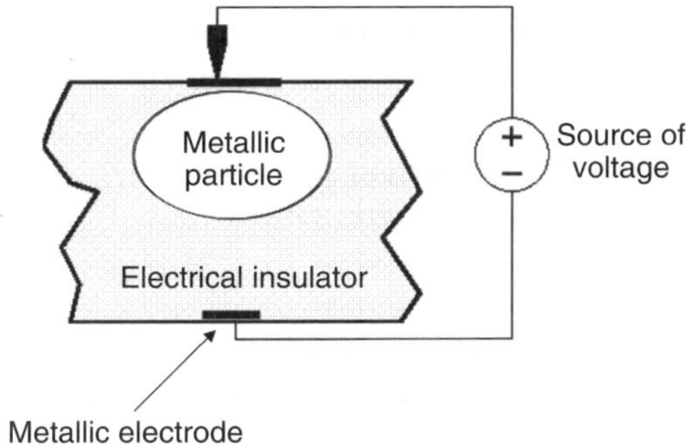

Figure 5.3. *The electron box. The metal particle is of nanometric dimensions, the distance between the superior electrode and the particle is also minute, allowing transfer of electrons by quantum tunneling*

This transfer can be made unidirectional by applying a small negative voltage to the superior electrode, pushing the electrons toward the metal particle and stopping them from going back to the superior electrode. The metal particle thus obtains an electric charge equal to that of a single electron.

Once this first charge is obtained, transferring a second electron becomes far more difficult. This is because the negative charge located in the metal particle exerts a strong electric field toward the outside, going against the transfer of a second electron.

This new barrier, known as the Coulomb Blockade[5], can still be overcome considering the electrons that have been transported by the

5 The electron that has been trapped in the particle acquires an amount of energy equal to $e^2/2C$, with e designating the electric charge of the electron, C the capacitance (an electrical value characteristic of capacitors) that varies according to the dimensions of the particle and its environment. The electric field caused by the excess electron in a 10-nm diameter particle is gigantic: it is equivalent to the electric field created between two metal plates 1 cm apart, and in between them there is a voltage of roughly 140,000 V. For the electron to remain trapped in the particle, its electrical energy must be greatly superior to its natural thermal energy, equal to kT (k: Boltzmann's constant, T: absolute temperature).

Thus, considering the example of an electron inside a 10-nm box, the electron's electrical energy (E_e) will be equal to roughly 100 meV (0.1 eV) and its thermal energy (E_T) increases by around 1/10 of a meV each time the temperature increases by 1 °C. The confinement of the electron inside the box will only be observable if E_T is a lot smaller than E_e. This requires the experiment to be held at very low temperatures, in the presence of liquid nitrogen, for example ($T = -196$ °C). Under these conditions, the thermal energy is lower than 10 meV, which is far inferior to the electrical energy. The same experiment can be done under higher temperature conditions, as long as the size of the particle is greatly reduced.

conductor to the base of the "wall" formed by the metal/insulator interface continue to accumulate. The accumulation of electrons causes an energy build up until it reaches that of the first electron located in the box. Once this level is reached, another electron can go through the insulator by quantum tunneling.

An image of quantum tunneling toward an "electron box" was first presented by Likharev, comparing the transfer of an electron through a metal/insulator junction to the dripping of water drop-by-drop from the extremity of a tube (Figure 5.4).

In this sense, one can consider that the hydraulic equivalent of the accumulation of an electric charge at the metal/insulator junction is the formation of a drop of water, slowly growing until it detaches itself from the extremity of a tube under the effects of its own weight.

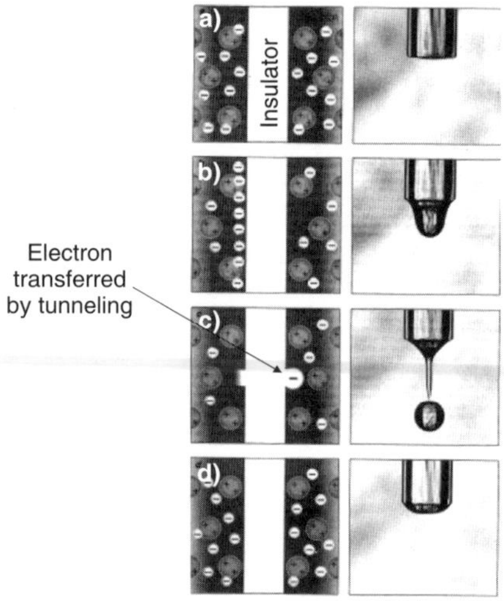

Figure 5.4. *Images comparing the transfer of an electron by quantum tunneling through an insulator and water dripping drop-by-drop from a tube. a) Equal charges on either side of the insulator. b) Electronic charges accumulate on the left under the effect of a potential difference applied between the two electrodes; the water drop appears and starts to grow. c) An electron has tunneled through the insulator, the amount of charge on the left is decreasing; the drop of water has become detached from the extremity of the tube. d) The process of electron accumulation is starting to repeat itself due to the potential applied to the electrode; a new drop of water starts to form. The bigger circles correspond to atoms and are immobile, the smaller ones represent electrons and are mobile (adapted from Likharev and Claeson [LIK 92])*

Once the second electron has been transferred to the metal particle, the balance of energy is once again inversed: the two excess electrons in the box exert an electric field that stops the transfer of a third one. A new accumulation has to take place to allow the transfer of a third electron, in the same way as a new drop of water starts to grow and then detach itself.

This is a periodical phenomenon and results in spikes of current at regular time intervals, which are directly related to the charge time of the electron box, not unlike that of a capacitor (Box 6.2)[6].

A similar phenomenon is observed if the potential at the inferior electrode is gradually increased. A series of intensity spikes is seen for increasing values of the electrode's potential. Each value is characteristic of the potential corresponding to an increment of energy transferred to the electron, and favorable to a transfer by quantum tunneling.

However, such an experiment requires certain particular conditions. The trapping of the electrons in the *island* is only possible if the electrons' natural thermal energy is vastly inferior to the electrical energy of the electron in the box. This means the experiment must be held at very low temperatures (the temperature of liquid nitrogen, for example) and the dimensions of the particle must be in the region of a few nanometers at the most.

It must also be remarked that, from a practical point of view, using this sort of device as part of an electronic circuit is not particularly useful. The number of electrons in the box is strictly dependent on the potential applied to the inferior electrode, and consequently it has no useful application in the storage of information. Moreover, such a system does not allow the measuring of a direct current.

5.2.1.2. *The single-electron transistor (SET)*

This works by merging the mode of operation of a standard transistor (like the MOSFET) with that of a device that allows continuous control of the passage of electrons. The information corresponds to the "ON" and "OFF"

---

6 By "charging" a capacitor with an electrical charge Q, its potential increases gradually as a function of time until reaching its maximal value V_L defined by its capacity C (itself defined by its geometrical characteristics) and by the resistance of the circuit R with $Q = C\ V_L$. The instantaneous potential V_t is given by the equation $V_t = V_L(1 - e^{-t/\tau})$; τ represents the capacitor's time constant, which means that when time $t = \tau$, the capacitor's potential is slightly greater than $V_L/2$.

functions of the transistor; in other words whether or not the current is flowing. In binary, this translates as either 1 or 0, respectively. The challenge is to be able to carry out these operations with a single electron when even the most advanced transistors need millions of electrons to do this.

The workings of such a device were first demonstrated by several teams in the late 1990s. Its creation, comparable to that of FET, is relatively complex. It consists of artificially creating an electron reservoir roughly 100 nm across. Like an FET, the reservoir is controlled by a gate electrode, and the current is measured in between two electrodes (source and drain).

Considering the experimental difficulties involved in the confinement of electrons using electrodes, other teams have considered using a nano-object like a semiconducting nanoparticle for the electron box.

Thus, a single-electron FET was created for the first time in 1997 by two Californian teams in Berkeley. The electron box is made up of a semiconducting nanoparticle of selenium sulfide covered in a single layer of molecules forming a very thin insulating barrier, which is necessary for the transfer of electrons by quantum tunneling. Experimentally, a considerable feat involves placing this nanoparticle in between two electrodes (source and drain), themselves only approximately 10 nm apart. The rest of the setup is the standard one used in the FET (Figure 5.5).

The resulting graphs confirm the general underlying principles. The currents measured as a function of the gate potential show values with spikes characteristic of Coulomb's Blockade. However, the differences in function between the "ON" and "OFF" positions are too minute to consider using them practically with modern technology.

Another method has been suggested, using carbon nanotubes (CNTs). Thanks to an international collaboration of physicists and chemists (The CEA in France, Delft University of Technology in the Netherlands, and Rice University in the United States), such an experiment was successfully carried out in 1997. Figure 5.6 shows a CNT with a diameter of 1 nm in contact with two platinum electrodes roughly 100 nm apart (source and drain).

As previously seen, one can assume that the electrons captured by the CNTs are distributed according to discrete values of energy. In the case of a CNT, these are very close together. These energy levels can be increased or decreased by varying the potential V_G of the gate electrode. Moreover, transfer of electrons into the CNT by tunneling can be controlled as a function of V_G. By operating under very low temperature conditions, a charge phenomenon similar to that seen in the electron box's Coulomb Blockade can be observed.

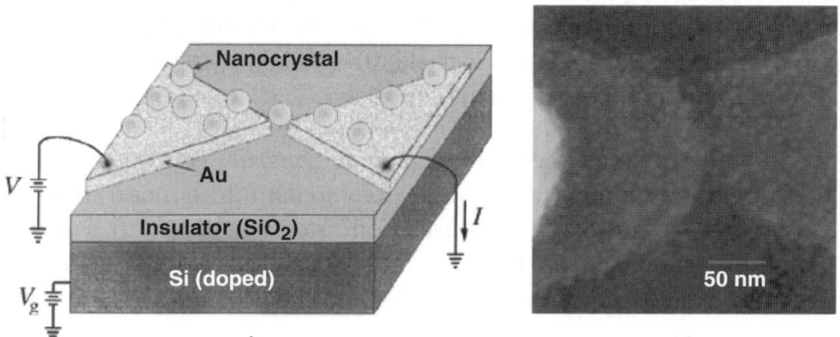

a) b)

Figure 5.5. *Single-electron transistor. a) Simplified diagram of a device similar to a FET: the insertion of a semiconducting NP or nanocrystal between the two gold electrodes 10 nm apart is the most difficult step in the making of the transistor. NPs are covered in a single layer of dithiolated insulating molecules (roughly 1-nm thick). These bind to the gold (Au) surface, surrounding the nanoparticle in a thin layer of insulation. b) SEM image of a junction. The dispersion of the NPs (diameter is about 5 nm) deposited on and between the two gold electrodes can be seen. The NPs are prepared in a colloidal solution and deposited by evaporation of the solvent (adapted from Klein et al. [KLE 97])*

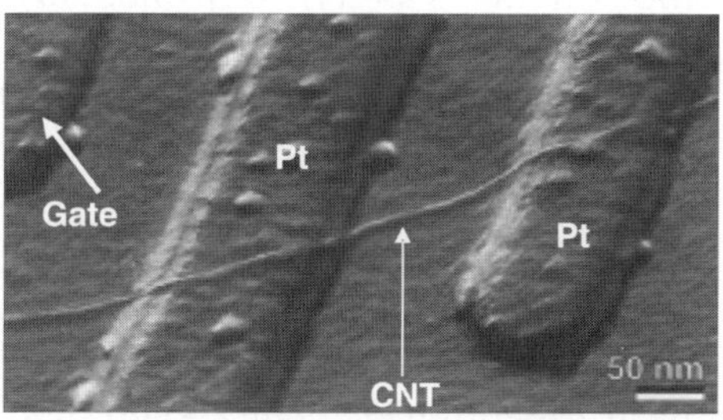

Figure 5.6. *SEM image of a CNT in contact with two platinum (Pt) electrodes. The electrodes were made by evaporation in a vacuum on a Si/SiO$_2$ substrate (an insulator). The gate on the left of the figure exerts an electric field on the CNT, modifying the energy levels available to the electrons (adapted from Tans et al. [TAN 97])*

By applying a low voltage (about a microvolt) to the two electrodes in contact with each other, and by varying the gate's electrical potential, a series of current spikes can be observed. These are characteristic of the charge phenomenon, and correspond to a transfer of electrons by tunneling (Figure 5.7).

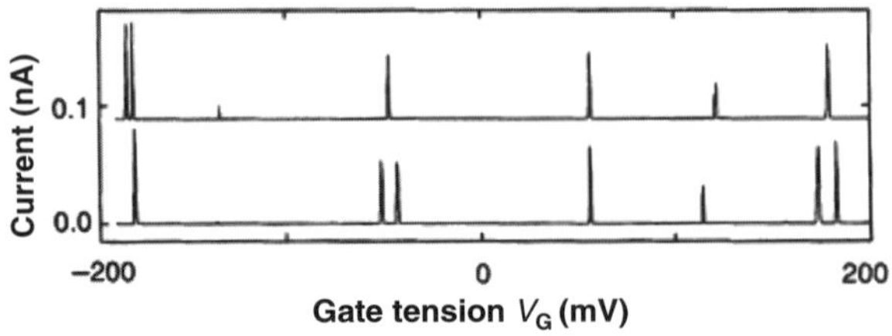

Figure 5.7. *Current spikes observed as a function of the gate potential V_G. The experiment was carried out twice, and confirms that distances between the spikes are the same each time. The double spikes observed are random and are caused by parasitic charges (adapted from Tans et al. [TAN 97])*

5.2.2. *The ultimate step: the molecule*

This constitutes an extra step in the miniaturization process. As opposed to previous systems that used nano-objects with sizes smaller than 10 nm, creating a transistor in which the only active element is a single molecule (the size of which can go from a tenth to several tenths of a nanometer) would theoretically allow a vast increase in the potential transistor density per unit of surface.

Moreover, using molecules instead of nano-objects would not only decrease the size of the component, more importantly it would also offer an infinite amount of different solutions and possibilities thanks to the electrical properties specific to different molecular families.

This, however, brings us right into the heart of quantum mechanics, creating new difficulties and challenges to overcome.

5.2.2.1. *Technical issues in the assembly of a metal/single molecule/metal junction*

In the 1970s, Aviram and Ratner [AVI 74] thought of a molecular device that behaved like a diode, rectifying current[7]. The idea took a long time to

7 A rectifier is an electronic device that can either block the electrical current in a circuit (OFF) or let it flow (ON) once a certain voltage is reached.

take physical shape, and it was only much later, in the 1990s, that the first concrete experiments started to take place.

There are several considerable problems to overcome. The major difficulty is establishing an electrical contact with the molecule. Two metallic electrodes are required to establish a flow of electrons in and out of the molecule (Figure 5.8).

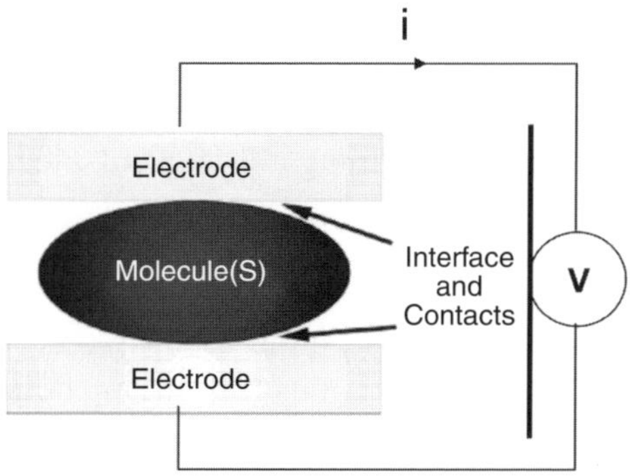

Figure 5.8. *Schematic representation of a metal/molecule/metal junction (adapted from Wang et al. [WAN 11a])*

Establishing these electrical connections is particularly tricky. It requires full mastery of all aspects of interface chemistry, profound knowledge in metal–molecule interactions, and very developed technical skills in using technologies involved in the thin film depositing of metals by evaporation. At this level, short circuits are very common between the two electrodes, often resulting in unwanted artifacts.

All these challenges might explain why, for the same molecule, different teams have had very different results. It is also for this reason that highly sophisticated techniques have been developed to avoid the direct deposit of the evaporation of a layer of metal (usually gold) on the molecule (Song *et al.*, 2011 [SON 11]).

The sheer complexity of the techniques involved in creating a nanospace between two electrodes, and then inserting a molecule in it, makes the use of this kind of device on a large scale impossible, and as a result, research into it has been provisionally discontinued. Instead, there has been a move toward systems involving chains of molecules, easier to control and closer to actual applications.

5.2.2.2. *Molecular diodes made from self-assembled organic molecules*

This allows the reduction of the amount of undesired artifacts and other random effects observed in the single-molecule device. Many molecular systems have been studied in the last few years, and already several devices nearly ready for large-scale manufacturing have been presented.

These techniques rely on the initial self-assembly of organic molecules, which have been previously adsorbed in an ordered manner on a metallic surface as a single layer of molecules (SAMs, or self-assembled molecules). After this monomolecular layer is formed, a thin layer of gold is deposited by evaporation in a vacuum, forming the second electrode in the metal/molecule/metal junction.

However, like previously, a major problem is that the atoms of gold obtained during the evaporation can quite easily pierce the molecular layer (which is thinner than a nanometer), thus forming metallic filaments that can cause short circuits and possible artifacts.

Several solutions have been suggested to overcome the difficulty of obtaining a reliable electrical contact between the metallic layer and the molecular one.

Several teams came up with the idea of first depositing a conductive intermediate layer that would not cause short circuits. This greatly improved reliability of the manufacturing process.

The latest solution consists of placing a sheet of graphene on the organic molecular layer, and then depositing a thin layer of gold by evaporation (Figure 5.9).

This process was only recently made possible by recent advances in the production of graphene. It presents many advantages compared to the old processes that used conductive layers of conductive polymers[8], or liquid metals with a low fusion point.

---

8 Unlike most classical polymers, which make very good electrical insulators, conductive polymers, discovered in the 1970s by MacDiarmid, Heeger and Shirakawa (who won Nobel Prize in 2000), make up a unique family of chemical compounds: in their neutral state they are insulators, but when they are oxidized they become excellent conductors. This property has led to widespread use in a variety of applications, especially in the manufacturing of different sensors (see Box 6.3).

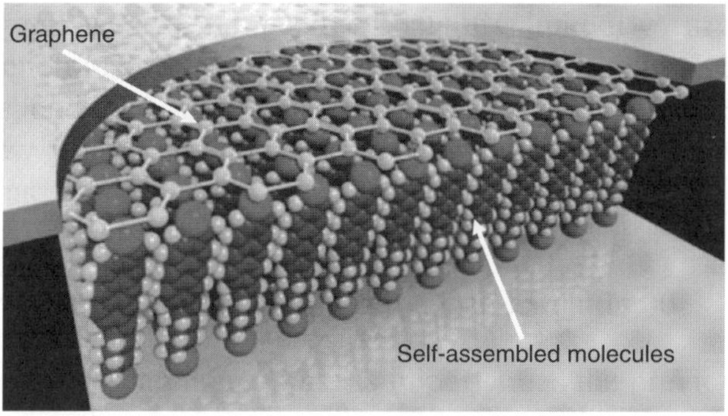

Figure 5.9. *Schematic representation of a micrometric layer of self-assembled SAMs. A sheet of graphene covers the top of the molecules. This layout is achieved by combining molecular self-assembly methods with classical lithography methods used in the microelectronics industry. The sheet of graphene, which is conductive, is then covered in a thin layer of gold, resulting in a diode [metal/organic molecules/graphene-metal] (adapted from Wang et al. [WAN 11b]). For a color version of this figure see www.iste.co.uk/lacaze/nano.zip*

5.2.2.3. *Electrical properties of self-assembled organic layers*

This is a subject that remains controversial from a theoretical point of view. A large number of different composites have been studied, yet the observations are very diverse. Nonetheless, two essential results can be extrapolated, differentiating two types of molecules: those with a passive behavior and those that react asymmetrically when an electric current flows through them. In most precise terms, this results in either symmetrical or asymmetrical electrical characteristics (intensity/voltage).

Symmetry of the electrical characteristics is seen in molecules that do not possess any redox[9] groups, and that are made exclusively from chains of hydrocarbons. An example of this is shown in Figure 5.10(a).

Molecules that have redox groups located at one end of the chain usually present asymmetrical characteristics. An example of this is shown in the molecules shown in Figure 5.10(b), which have a redox ferrocene group at their extremity.

---

9 A redox function is a chemical group (molecule, ion) that is likely to exchange electrons with another chemical group, a phenomenon that is called a reduction-oxidation reaction. These electronic exchanges between two chemical groups are characteristic of redox couples. For example, Fe^{2+} and Fe^{3+} (ferrous and ferric ions) constitute a redox couple; Fe^{2+} is oxidized to Fe^{3+} with the loss of an electron. The inverse reaction is a reduction and involves the gain of an electron (see Box 6.2, 159).

This asymmetry is important as it is this property that can potentially be used for rectifying or memory effects, which can be used in digital devices.

This rectification effect, characterized by a ratio of the different intensities of the current running through the molecular diode between the ON and OFF states, is a lot weaker than that obtained in the classical FET devices found in modern processors, which is in the order of one million (a minimum of 100,000 is needed for a logic circuit to work).

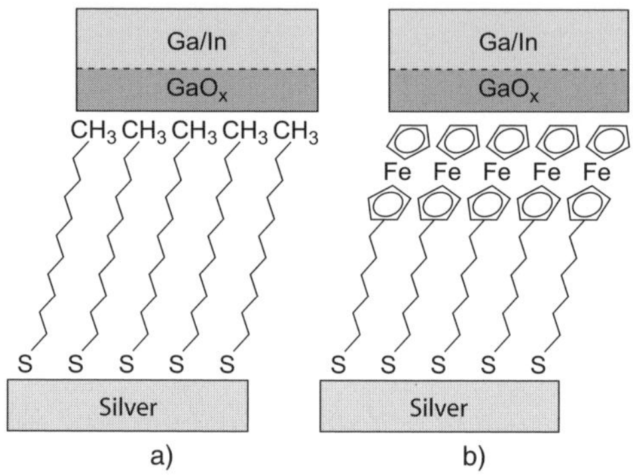

Figure 5.10. *Molecular diodes made of self-assembled molecules on a silver surface, covered in a conducting intermediate layer of gallium indium alloy (characterized by a low fusion point). The sulfur atom at the end of the chain guarantees a strong bond between the molecule and the silver. a) Molecules without a redox group; b) molecules with a ferrocene redox group (Fe^{2+} ion bound to two negative cyclopentadienyl ions) (adapted from Nijhuis et al. [NIJ 10])*

The electrical characteristics (Figure 5.11) obtained using the asymmetrical molecules of Figure 5.10 do confirm this.

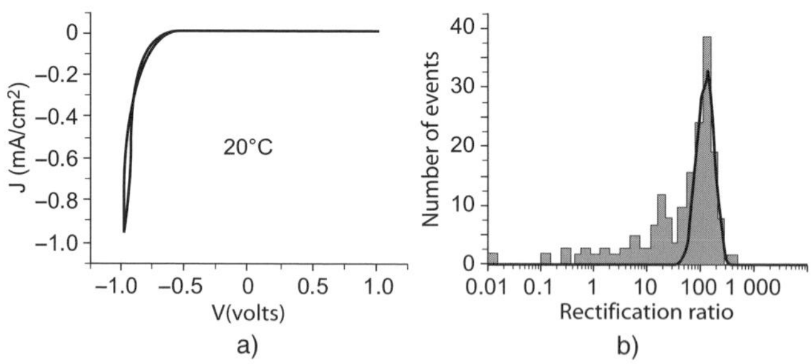

Figure 5.11. *a) Typical intensity/potential curve in the case of asymmetrical HS-(CH_2)n-Ferrocene molecules. The rectification ratio is in the order of 100 and corresponds to the ratio J(–1 V)/J(+1 V). b) Histogram of the rectification ratios, showing that the reproducibility of the experiments is not 100% (adapted from Nijhuis et al. [NIJ 10])*

5.2.2.4. *The organic field-effect transistor*

The idea is to replace an inorganic MOSFET with an entirely plastic field-effect transistor, called an organic field-effect transistor (OFET). The advantages over a traditional MOSFET would include greater component flexibility, reduced weight and reduced cost to produce, allowing potential large-scale production at a lower cost. These would be particularly suitable for use in huge display panels.

It is this possible application in particular that has driven a lot of the interest in the field, and led to it being given the name organic and large area electronics, or OLAE (Mas-Torrent *et al.* [MAS 11]).

In reality though, such an objective is still far away, and even the most recent studies have mainly focused on FETs in which the active layer of the semiconductor is an organic semiconductor instead of silicon.

Two main families of organic compounds have been tested: polymers and molecules with a low molecular weight, characterized by the presence of conjugated unsaturated bonds (alternating double and single bonds) that increase the movement of electrons within the molecule, necessary for obtaining good electron mobility in the material.

A lot of work has been done using pentacene as an organic semiconductor. Pentacene is a molecule consisting of a condensed chain of five benzene rings, and can be assimilated to a fragment of graphene (Figure 5.12). Pentacene is solid at room temperature, and is usually deposited by sublimation[10] and condensation on an insulating surface. In this way, self-assembled monomolecular layers roughly 20-nm thick can be obtained.

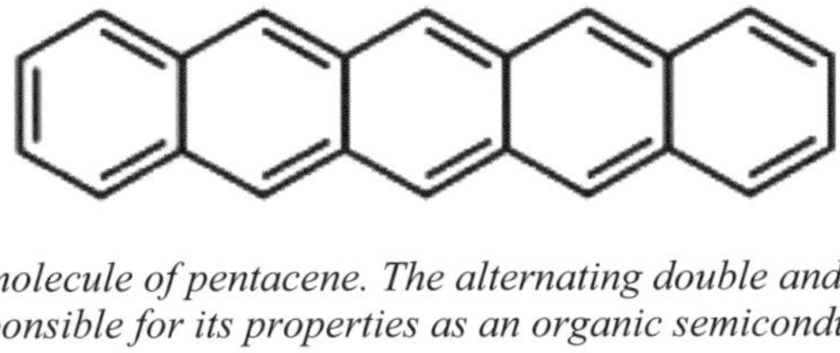

Figure 5.12. *A molecule of pentacene. The alternating double and single bonds are responsible for its properties as an organic semiconductor*

10 Sublimation is the evaporation in a vacuum of solid compounds, characterized by a strong vapor tension. The deposit is obtained by condensation of the vapor on a solid surface at a low temperature.

Of the numerous problems presented by these systems, two are particularly important: (1) the mobility of the charge carriers in the organic active layer is usually very limited, which requires the application of high voltages; (2) the cut-in voltage necessary to move the transistor from the ON to the OFF position is sometimes poorly defined.

The mobility of the charge carriers is an electrical characteristic dependent on the nature of the organic material. It defines the speed of the charges as they move around the material under the influence of an electric field, and can vary by several orders of magnitude. By taking for reference the generally accepted value for amorphous silicon (1 cm^2/V.s), materials that are not very conductive can have values 10,000 times smaller, organic materials with medium levels of structural order 100 times smaller and organic materials with a crystalline level of order have similar or higher levels of mobility than amorphous silicon[11].

Like before, these organic layers must be flawless. Thus, obtaining these consistently is not an easy task. This, however, is no way near as big a problem as those encountered in the production of two-electrode molecular diodes, as the pentacene is deposited on an insulating layer, which eliminates any risk of a short circuit between the electrodes and the gate (Figure 5.13).

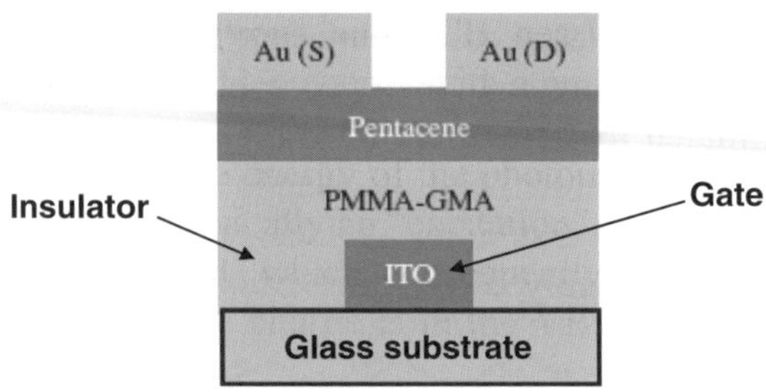

Figure 5.13. *Pentacene-based OFET. The gold (Au), source (S) and drain (D) electrodes are deposited at the extremities of the pentacene layer (thickness of 20 nm) by evaporation. The gate electrode is an indium tin oxide (ITO) deposit, on which a layer (400 nm thick) of an insulating polymer (PMMA-GMA or polymethyl methacrylate-glycidyl methacrylate) has been deposited. In this layout, pentacene's active zone is in the open air (adapted from Wang et al. [WAN 07])*

11 Crystalline silicon with a high level of purity (very expensive compared to amorphous silicon) has mobility levels 1,000 times higher than that of amorphous silicon.

The electrical characteristics are analogous to those of a classical MOSFET. However, the tension required for correct function remains too high for widespread use. Current research is focusing mainly on increasing the mobility of the charges in the organic layer and discovering new molecules and polymers with better semiconducting properties than amorphous silicon.

5.2.3. *Conclusion*

Molecular electronics are still a domain of research where huge progress needs to be made before production on a large scale can be envisaged. A number of problems in relation to reproducibility, the choice of molecular systems, their assembly on flawless surfaces, etc. are issues that are still to be fully overcome.

Understanding of the electronic properties of molecules has indeed made some considerable progress, but there are many points that remain obscure, particularly in relation to the role of metal–molecule contacts. This is a critical point, especially in layouts involving molecular diodes (two electrodes), where the rectification ratio remains too low (a few hundred) to be able to consider their application in logic circuits, which require rectification ratios at least in the order of 100,000.

Production of OFETs on the other hand seems a bit more feasible. They could soon enough give rise to applications in plastic electronics, or in the large-scale, low-cost production of OFET transistors. This, however, depends upon the discovery of molecular systems with a level of electrical mobility similar to that of crystalline silicon.

5.3. Spintronics

Traditional electronics are founded on the assumption that electrons, carrying identical electrical charges, have identical conducting properties when submitted to the actions on an electrical field. This is true as long as the electrons are traveling through conductive or semiconducting non-magnetic materials.

In reality, electrons are not equivalent. They differ fundamentally in their magnetic properties, and these distinct magnetic properties can be revealed by the application of a magnetic field that are at the origin of a new kind of electronics, called spin electronics, or spintronics.

5.3.1. *Electron spin and ferromagnetic materials*

Electrons can be visualized as miniscule tops spinning on themselves according to two opposite directions. This rotational movement (electron spin) gives the electrons magnetic properties, making them comparable to tiny magnets (Figure 5.14). There are two possible orientations, giving rise to two distinct spin populations: magnetic moment directed upward (spin "up" ↑) and downward (spin "down" ↓).

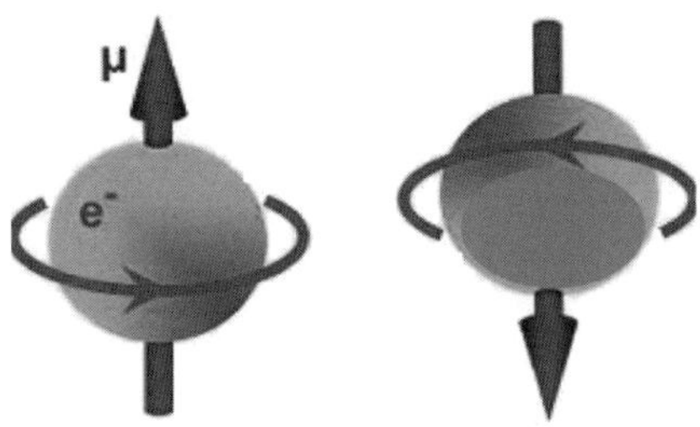

Figure 5.14. *Electron spin. The electron, visualized as an electrically charged sphere, spinning on itself, which results in it having the magnetic properties of a magnet, characterized by a magnetic moment* μ *(a magnetic moment defines the degree of magnetization). This magnet can have two opposite orientations, which correspond to the clockwise and anticlockwise rotations the electron can have on itself*

The conductive properties of these two types of electrons, identical in traditional electronics, diverge when they are located in a ferromagnetic conductor like iron, nickel, cobalt or an alloy like Fe-Ni.

Why this difference? The reason is ferromagnetic conductors themselves are very unique materials.

Although most molecular materials are formed by covalent bonds in which the electrons are associated in pairs of opposite spin, compensating for each other, iron is characterized by atoms with isolated electrons (not in pairs). These atoms form microdomains called magnetic domains inside which the electrons' spins are orientated in the same direction, but different from one domain to the other.

In the absence of a magnetic field, the global magnetization of a ferromagnetic material is nil (Figure 5.15(a)). However, when an external H_1 and H_2 magnetic field is applied, this is no longer the case.

According to the intensity of this magnetic field, a progressive reorientation of the spin microdomains takes place (Figure 5.15(b)), and ultimately, if the magnetic field has a very high intensity, all the spins of all the electrons are orientated so as to be parallel to the applied field *H* (Figure 5.15(c)).

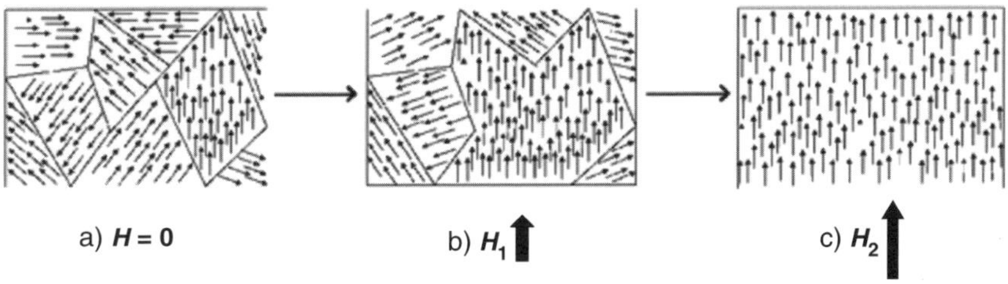

Figure 5.15. *Evolution of the magnetization of a ferromagnetic metal under the action of a magnetic field H. a) Magnetic microdomains in the absence of a magnetic field; b) the fusion of three magnetic domains and alignment of the spins inside the new domain and progressive reorientation in the other domains under the effect of the magnetic field H_1; c) fusion of all the domains caused by the magnetic field H_2, more intense than H_1, and full alignment of the spin moments so as to be parallel to H*

At this point, the metal has acquired a certain degree of magnetization, which partially disappears when the magnetic field is interrupted. The material does, however, keep a residual magnetization (called remanent magnetization) of varying strength that can be used in the production of electromagnets (soft iron) or permanent magnets if the remanent magnetization is very high (ferrite, iron oxide of the formula Fe_3O_4).

5.3.2. *Magnetoresistance*

All materials do not conduct an electric current in the same way, and actually block its passage to a certain extent. Each material is characterized by its electrical resistance; a measurement that defines how easy electron transport is within the material. Metals, which are very good conductors, are characterized by a near-nil electrical resistance, whereas, on the other hand, insulators, totally opposed to the flow of current, have extremely high electrical resistances.

In the case of ferromagnetic materials, another distinction can be made, directly relating the spin of electrons. Using the example of some iron in the presence of a magnetic field, it can be observed that the material's electrical resistance is higher than the resistance measured in the absence of a magnetic field. This is magnetoresistance (MR), discovered by Lord Rayleigh more than a century ago.

Why this effect? The answer is simple enough, and comes down to assuming that electrons with ''up'' ↑ and ''down'' ↓ spins will undergo different "frictions" depending on whether they come into contact with same (parallel spins ↑↑) or different (antiparallel spins ↑↓) spin orientations. This comes down to saying that the overall current is the superposition of two currents that propagate throughout the metal while undergoing different interactions. As it turns out, the "friction" between parallel spins is weaker than that between anti-parallel spins, meaning that electrical resistance is higher when the spins are antiparallel.

This is exactly what happens in a ferromagnetic material like iron under the influence of a magnetic field. The iron atoms' electron spins become orientated according to the direction of the applied field, whereas the electrons responsible for the electrical current possess (statistically) two spin orientations. As a result, half of the current is carried by electrons with a spin parallel to that of the iron atoms' electrons, whereas the other half is carried by electrons with an opposite spin to that of the iron atoms' electrons. Globally, this causes electrical resistance variations when the intensity of the magnetic field is increased: this is the effect of MR. It was described in the 1970s and corresponds to relative resistance variation in the order of 2%.

5.3.3. *Giant magnetoresistance*

Giant magnetoresistance (GMR) is a new property of MR, calling upon nanotechnology, as the effect was observed in the case of alternating magnetic and non-magnetic ultrathin layers (with a thickness of few atoms, 2–3 nm).

When the electrical current travels through multiple ultrathin layers of ferromagnetic metals exposed to a magnetic field, the MR decreases considerably and the variation in resistance is no longer 2–5%, but rather in the order of 80–100%. This variation is huge, hence the name giant magnetoresistance.

The phenomenon was first observed in 1988 by Albert Fert in the Laboratoire de Physique des Solides d'Orsay in collaboration with the Laboratoire Central de Recherche Thomson CSF [BAI 88]. Peter Grünberg was making the same discovery in Germany, but had the foresight to patent the invention before publishing. Both received the Nobel Prize in physics in 2007.

The 1988 publication describes the effect in a system with ultrathin, superimposed iron/chromium/iron layers; the chromium layer (non-ferromagnetic) playing the role of a separator between the iron ferromagnetic layers (Figure 5.16).

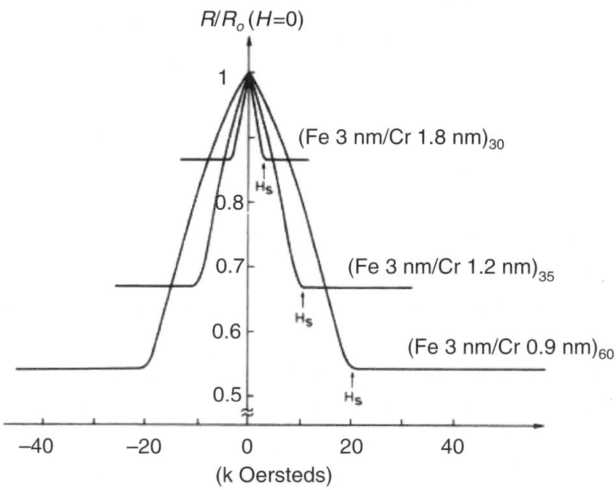

Figure 5.16. *Relative variation in magnetoresistance as a function of the field applied (measure in Oersteds) for three networks of multiple alternating Fe/Cr layers at very low temperature. The indices 30, 35 and 60 represent the number of alternating Fe/Cr layers (adapted from Bai et al. [BAI 88])*

The variation in MR increases as the thickness of the chromium layer decreases. It increases with the field applied, until reaching a limit when the field becomes greater than Hs, which corresponds to a saturation of the magnetization in the iron layers.

The flow of current through these layers, as represented in Figure 5.17, is facilitated when the electrons have a parallel spin to those of the material (little resistance, straight arrow). Conversely, when the spins have opposite orientations, the material's resistivity is greater (broken arrow).

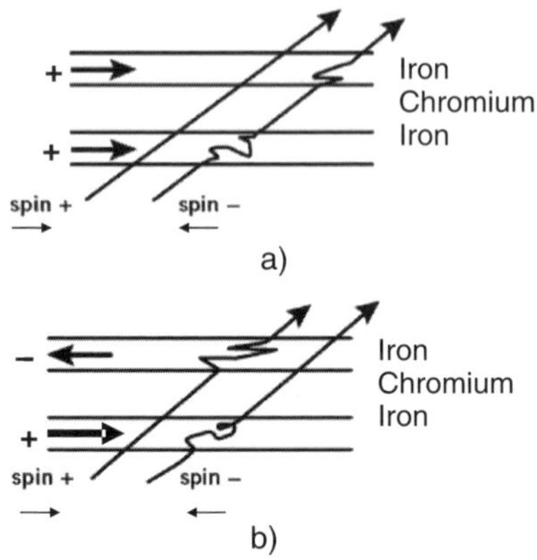

Figure 5.17. *Simplified diagram of three ultrathin superimposed iron/chromium/iron layers, characterized by a) a parallel magnetization of the iron layers (marked +, +) and b) an antiparallel magnetization of the iron layers (marked –, +). The chromium layer is not ferromagnetic and functions as a magnetic insulation layer between the two iron layers. The arrows going through the alternating iron/chromium/iron layers represent the two channels of electrons carried by the current of antiparallel spins, marked + and –. The straight arrows indicates that the electrons go through the layer without resistance, as opposed to the broken arrows that indicate the existence of resistance (adapted from an article published in the journal Plein Sud (Orsay), January–February 2008, p. 11)*

In these conditions, by considering that the current can be divided into two electron "channels" of opposite + and – spin, it can be observed in the case where the magnetic layers have parallel spins that the electrons of one of the two electron channels (spins parallel to those of the magnetic layers) go through the layers without resistance (equivalent to a short circuit), whereas the electrons from the second channel (spins antiparallel to the magnetic layers) are slowed (broken arrow), Figure 5.17(a).

Although the magnetic layers have their spins antiparallel, Figure 5.17(b), the electrons from each of the channels have to come in contact with a magnetic layer with an antiparallel spin, causing a globally higher resistance than before (no short-circuited electron channel).

To conclude, the parallel orientation of the spins of the different magnetic layers means that one of the electron channels is short-circuited, whereas in the case of an antiparallel orientation, the two channels are resistant. This explains the decrease in MR when all the ferromagnetic layers present the same spin orientation.

This is an important discovery, and plays a crucial role in the operating of the magnetic heads associated with the hard drives of modern computers. The fact that the variation in resistance is considerably increased when the magnetic dots are made of alternating "ferromagnetic material/non-ferromagnetic material" multilayers allows a considerable increase in detection sensitivity, and consequently enables an equal decrease in the corresponding dots' size, which has resulted in an increase in the storage capacity of hard drives by about two orders of magnitude, without which laptop computers would not have been possible.

Spintronics is a relatively young field that joins magnetism and electronics. Since the discovery of GMR, spintronics has given rise to new developments such as the conception of devices like the magnetoresistance tunnel (MRT), in which the non-magnetic metallic layers are replaced by insulating layers that the electrons go through by quantum tunneling.

In this case, the variation in resistance is much greater than that of the classic GMR. Since the 1990s when GMR was first introduced into the magnetic heads, the storage space available has increased by three orders of magnitude, reaching storage densities of roughly 30 Gb/cm^2 (the theoretical limit being currently estimated at 160 Gb/cm^2) (Nozières [NOZ 10]).

5.4. Conclusions

As we were able to observe, industrial electronics has long since passed from the domain of microelectronics to the domain of nanoelectronics. Top-down strategies, still used today, reached the scale of the nanometer two decades ago.

The very rapid evolution of deposition techniques under a vacuum of material layers, which constitutes the basis of planar technology, has allowed us to reach fantastic densities of electronic components. The reliability concerning the production of these components is also exceptional. Surface observation methods are becoming increasingly powerful and knowledge in the field of the physicochemistry of ultrathin layer materials is increasing incessantly, resulting in the optimized quality of the deposits.

All the predictions, however, seem to agree on a limit to the progression of miniaturization. Other solutions will also have to be found to reduce the energy consumption of computers.

An ideal solution could possibly come in the form of molecular electronics, often hailed as the electronics of the future. However, it remains far away, both from a research point of view, and even more so from a commercial point of view, as it would require automated molecular layer self-assembly processes.

Spintronics is a science in full evolution, and has already provided its share of applications. The recent observation of GMR in graphene nano-ribbons might mark the start of a new direction in research, which could lead in the near future to very important discoveries.

5.5. Bibliography

[AVI 74] AVIRAM A., RATNER M.A., "Molecular rectifiers", *Chemical Physics Letters*, vol. 29, pp. 277–283, 1974.

[BAI 88] BAIBICH M.N., BROTO J.M., FERT A., NGUYEN VAN DAU F., PETROFF F., ETIENNE P., CREUZET G., FRIEDERICH A., CHAZELAS J., "Giant magnetoresistance of (001)Fe/(001)Cr magnetic superlattices", *Journal of Physics Review Letters*, vol. 61, pp. 2472–2475, 1988.

[KLE 97] KLEIN D.L., ROTH R., LIM A.K.L., ALIVISATOS A.P., MCEUEN P.L., "A single-electron transistor made from a cadmium selenide nanocrystal", *Nature*, vol. 389, pp. 699–701, 1997.

[LIK 92] LIKHAREV K.K., CLAESON T., "Single electronics", *Scientific American*, vol. 266, pp. 80–85, 1992.

[LIK 99] LIKHAREV K.K., "Single-electron devices and their applications", *Proceedings of the IEEE*, vol. 87, pp. 606–632, 1999.

[MAS 11] MAS-TORRENT M., ROVIRA C., "The role of molecular order and solid-state structure in organic field-effect transistors (OFETs)", *Chemical Reviews*, vol. 111, pp. 4833–4856, 2011.

[NIJ 10] NIJHUIS C.A., REUS W.F., BARBER J.R., DICKEY M.D., WHITESIDES G.M., "Charge transport and rectification in arrays of SAM-based tunneling junctions", *Nano Letters*, vol. 10, pp. 3611–3619, 2010.

[NOZ 10] NOZIERES J.-P., "Têtes de lecture et mémoires magnétiques", *Reflets de la Physique, Société Française de Physique, Dossier Spintronique*, pp. 34–38, 2010, available at www.refletsdelaphysique.fr

[SON 11] SONG H., REED M.A., LEE T., "Single molecule electronic devices", *Advanced Materials*, vol. 23, pp. 1583–1608, 2011.

[TAN 97] TANS S.J., DEVORET M.H., DAI H., THESS A., SMALLEY R.E., GEERLIGS L.J., DEKER C., "Individual single-wall carbon nanotubes as quantum wires", *Nature*, vol. 386, pp. 474–477, 1997.

[WAN 07] WANG W., SHI J., JIANG W., GUO S., ZHANG H., QUAN B., MA D., "High mobility pentacene thin-film transistors with copolymer-gate dielectrics", *Microelectronics Journal*, vol. 38, pp. 27–30, 2007.

[WAN 11a] WANG G., KIM T.-W., LEE T., "Electrical transport characteristics through molecular layers", *Journal of Materials Chemistry*, vol. 21, pp. 18117–18136, 2011.

[WAN 11b] WANG G., KIM Y., CHOE M., KIM T.-W., LEE T., "A new approach for molecular electronic junctions with a multilayer graphene electrode", *Advanced Materials*, vol. 23, pp. 755–760, 2011.

Chapter 6

Energy and Nanomaterials

Energy has always been a major preoccupation of modern societies. In the 21st Century, this is all the more true with a new focus on protecting the environment.

The European Union's FP7 (7th Framework Program for research and technological development), with its 50 billion euro budget, places a lot of importance on energy. The presentation of the strategic energy technology (SET) plan, which incites industries to use energy-efficient technologies with low carbon emissions, or in other words which incites them to invest in and develop renewable energy sources, is the direct result of this.

It is precisely in the area of solar energy and low carbon-output energy storage where nanoscience has a very important role to play; a role in which physics and chemistry work together and innovation requires the help of nanomaterials.

The storage of electrical energy electrochemically, which implies a type of carbon-free energy production, is a rare domain in industrial production that is in near-constant growth and progression.

The multitude of projects involving all-electric cars currently being undertaken and the increase in the number of mobile electronic devices should guarantee growth in the sector for many more years to come.

Moreover, the industry involved in the conversion of solar energy into electrical energy is booming. Although it is true that silicon-based technology is well established and present in the majority of equipment, important research continues to develop new systems based on nanomaterials, as efficient as silicon but a lot less expensive.

The use of organic, low-cost materials that can be applied on to large surfaces like paint is a promising field, providing performances that are continually improving and getting closer and closer to silicon.

6.1. Electrochemical storage of electricity

This is a major problem affecting not only the moderation of domestic and industrial electrical consumption from distribution networks but also the various electrical equipment used in everyday life (portable system batteries, storage devices for transient security systems, accumulators for the electrical equipment of thermal engine vehicles, etc.).

The theory of an electric battery is simple and relies on the conversion of chemical energy into electrical energy, so-called an electrochemical reaction (Box 6.1).

The advantage of such a system is that the conversion of chemical energy into electrical energy is done with a yield close to 100%.

Conceptually, the system is the same as the one used in the Volta[1] battery, invented over two centuries ago. It consists of a piled up series of copper and zinc discs, each one separated from the other by some cloth soaked in salt water, able to supply an electric current between its two extremities (Figure 6.1).

Since Volta's invention, a large number of systems have been put forward, but only a small number of them have been used commercially or in major applications.

---

1 Alessandro Volta (1745–1827) was an Italian physicist, famous for inventing the first battery in 1800. His invention quickly traveled the length of Europe, and Napoleon himself came to witness Volta's presentation of the invention in front of the members of the Institute of France. The invention was a huge step in the history of chemistry, and was decisive in the discovery of a number of other elements. Humphry Davy (1778–1829) was the first, in the early 1800s, to isolate several metals (notably sodium, potassium, magnesium, calcium, strontium, barium) by electrolyzing salts using the Volta battery.

Box 6.1. Electrodes and electrochemical cells

Redox potential of an electrode

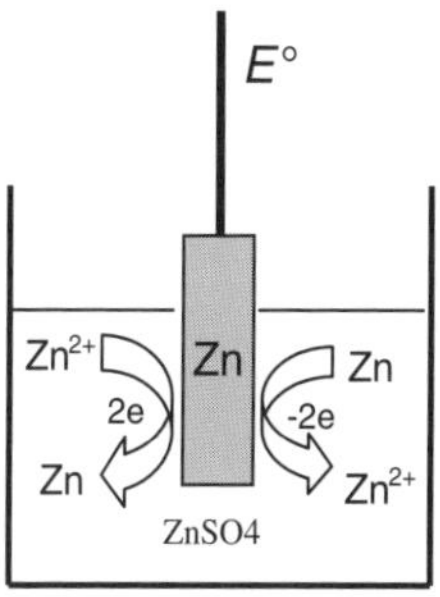

A zinc electrode immersed in a solution of zinc sulfate, dissociated into Zn^{2+} and SO_4^{2-}, spontaneously acquires an equilibrium potential $E^{\circ}_{Zn^{2+}/Zn}$, (redox potential), which corresponds to the exchange of two electrons between Zn^{2+} (Ox) and Zn (Red).

If the applied potential E is greater than $E^{\circ}_{Zn^{2+}/Zn}$, the electronic exchange takes place in the direction $Zn \rightarrow Zn^{2+} + 2e^-$, and the zinc is oxidized (the electrode is an anode).

$E < E°(Zn^{2+}/Zn)$ $E > E°(Zn^{2+}/Zn)$

In the opposite scenario, with E less than $E^{\circ}_{Zn^{2+}/Zn}$, the exchange reaction is $Zn^{2+} + 2e^- \rightarrow Zn$, with the cation Zn^{2+} being reduced (the electrode is a cathode).

Galvanic cell (battery)

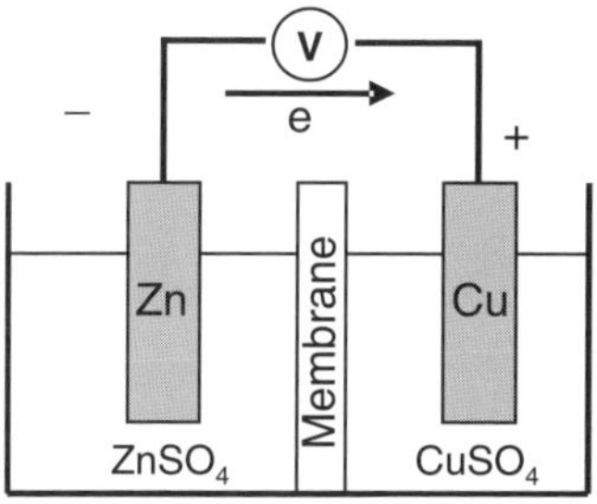

Two electrodes (zinc and copper) are dipped into two compartments ($ZnSO_4$ and $CuSO_4$) that are separated by a porous membrane, permeable to ions. The whole setup is a battery with an electromotive force V. The copper electrode is positive; the spontaneous reaction $Cu^{2+} + Zn \rightarrow Cu^{o} + Zn^{2+}$ corresponds to the transfer of electrons from the negative electrode toward the positive electrode. The zinc is dissolved, whereas the copper gets thicker.

Electrolysis cell

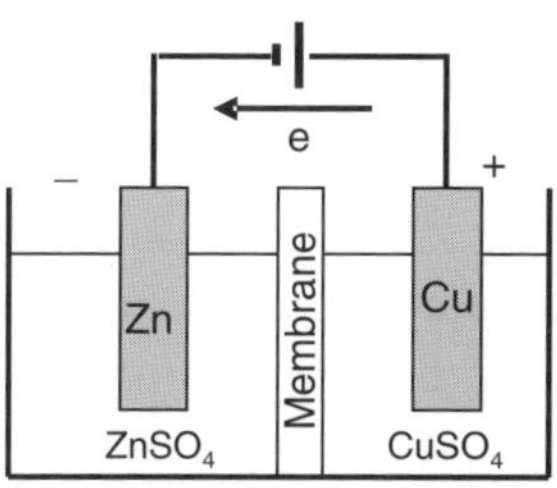

The setup is the same as before, except that tension is applied so as to "force" the inverse reaction, $Zn^{2+} + Cu \rightarrow Zn + Cu^{2+}$. The electrons move in the opposite direction to the previous direction. In this case, the copper is dissolved and the zinc gets thicker.

Lead accumulators (rechargeable batteries), invented by Gaston Planté in 1859, are still used today in the electrical equipment of thermal engine vehicules. Sodium–sulfur accumulators, more recent, were commercialized to build large storage units.

Figure 6.1. *Zinc–copper battery invented in 1800*

6.1.1. *Electrical properties of an accumulator*

An accumulator is a system that allows the storage of a finite amount of electrical energy, and can return it upon demand until the stock runs out. Weight and volume are the two main limitations, and they restrict the use of these devices to certain functions.

For systems with low electricity consumptions, like most portable electronic devices, small rechargeable batteries are sufficient for several hours of autonomous use.

However, an all-electric motor vehicle, for example, is going to be required to carry a high weight and/or a large volume, which in this sense are the limiting factors in the vehicle's autonomy.

These considerations have resulted in the defining of several energy-related units that take into consideration weight (specific energy and specific power) and volume (energy density). These units allow the expression of a battery's autonomy for a given weight or volume. Specific power is an equally important property as it defines the speed at which the energy can be delivered.

In practice, the commonly used units are the kilowatt-hour per kg (kWh/kg), defining a quantity of energy per unit of mass, the kilowatt-hour per liter (kWh/l)[2], energy per unit of volume, and finally the kilowatt per kg (kW/kg) that expresses the speed at which the energy can be delivered.

On top of these, there is also the criterion of rechargeability, that is the number of battery cycles (charges and discharge) a system can go through without degradation of the electrical properties.

Figure 6.2 gives an idea of the energies and specific powers of a few rechargeable battery systems. It can be noted that in these classifications, the Li-ion batteries give the best performances, which explains why these batteries tend to become widespread in most of the sectors where storing energy is necessary.

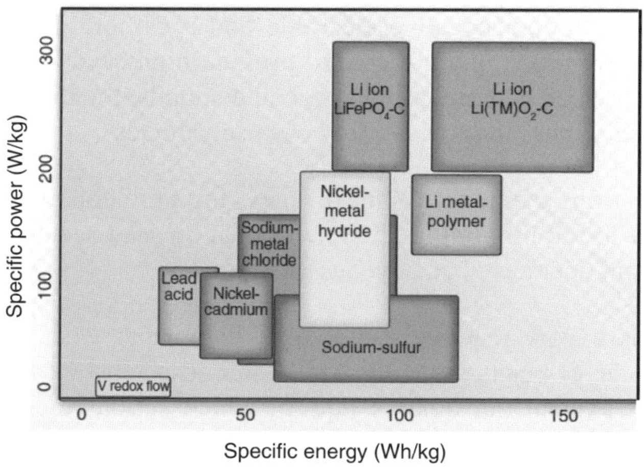

Figure 6.2. *Comparison of different rechargeable systems based on their recently established specific energies and specific powers (adapted from Dunn et al. [DUN 11])*

6.1.2. *Lithium batteries*

Figure 6.2 clearly shows that for all-electric vehicles the lithium-based electrochemical system appears as the best suited system to supply electrical energy with the best conditions.

2 A kWh is defined as 1,000 times the amount of energy consumed in 1 h by a current of 1 A with a voltage of 1 V. This is the energy equivalent of 1,000 times 3,600 J (the joule is a unit of energy, the watt a unit of power corresponding to a consumption of 1 J/s).

The reasons are simple: not only is lithium the lightest metal but it is also the metal that allows the highest voltages, resulting into more energy for less current used.

Lithium has been used as a metal electrode for quite a long time. The first lithium-based accumulators appeared in the 1970s, and were mainly built for applications in the military (torpedo motors for example).

Many systems, characterized by a negative metallic lithium electrode coupled with a positive electrode made of a variety of materials such as carbon monofluoride (CF_x), manganese dioxide (MnO_2), iron sulfur (FeS_2) and copper sulfur (CuS), have been suggested.

Moreover, Lithium-Oxygen ($Li\text{-}O_2$) and Lithium-Sulfur (Li-S) systems that also use a lithium negative electrode and a carbon positive electrode have many advantages, and are able to provide high specific energies. The $Li\text{-}O_2$ system, which is more promising, could soon be the much sought-after solution for equipping long-range fully electric vehicles[3].

However, from a practical point of view, the problems to overcome are much trickier for all lithium-based systems than for lead accumulators, or for any system that operates in an aqueous medium.

Lithium is a very reactive metal. It is flammable when it comes into contact with air. It reacts violently with water so it must be kept in contact with an inert solvent (an organic liquid), which seriously complicates the technology and the practical applications.

Much of the progress made, essentially of a technological nature, is mainly regarding the choice of solvents and electrolytes, or the optimization of the various electrode materials, which must be made more compatible with the limitations of electrochemical reactions. It is a complex domain in which performance has greatly improved because of the combined efforts of electrochemists and specialists in solids chemistry.

---

3 The different problems involving electrical energy storage are extensively covered in various reports and review articles. See, for example, Dunn *et al.* [DUN 11], Bruce *et al.* [BRU 12], Winter and Brodd [WIN 04], 2010 report from the Electric Power Research Institute (EPRI) [EPR 10], 2007 report from the Department of Energy (DOE-USA) [DOE 07].

An important technological breakthrough took place in the 1990s with the appearance of rechargeable Li-ion batteries, first created by the Japanese duo Nagaura and Tozawa [NAG 90].

The concept of these rechargeable batteries is radically different from that of the previous lithium batteries. It is in fact a process that involves the reversible insertion of Li^+ ions into two different electrode materials that creates the battery's very high voltage (Li^+ is a lithium atom that has lost an electron).

The advantages over metal lithium are significant: (1) the system is reversible and rechargeable; (2) it is safer to use as the lithium is no longer in its metallic state, but rather in the inert form of the Li^+ ion; (3) the battery life and the charge time are very long; (4) the maximal voltage E supplied by the battery is very high and can reach 3.8 V (Figure 6.3).

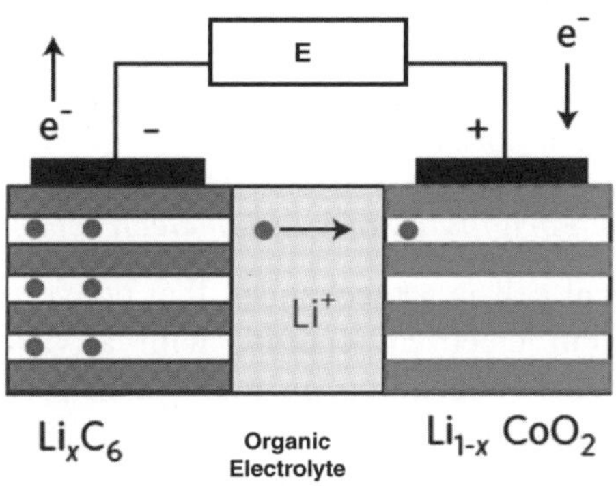

Figure 6.3. *Conceptual diagram of a Li-ion accumulator during discharge. The two electrodes are lamellar materials made of carbon (C) and cobalt oxide (CoO_2) in which lithium ions (Li^+) insert themselves. The discharge reaction equates to transferring the Li^+ ions from the carbon electrode to the cobalt electrode, which corresponds to the overall reaction: $0.5\ C_6Li + Li_{0.5}CoO_2 \rightarrow 3\ C + LiCoO_2$. In the state of intermediate discharge, the amount of lithium in each material is equal to Li_xC_6 and $Li_{1-x}CoO_2$ (x is a value between 0 and 1) (adapted from Bruce and Tarascon [BRU 12])*

Commercial success was immediate. Sony in Japan was the first to equip its electronic devices with this type of battery. Production increased dramatically; the number of Li-ion cells produced between 1993 and 2000 went from practically 0 to nearly 700 million units Brodd *et al.*, [BRO 00] and [BRO 04].

Use of the battery in fully electric cars is more recent. It allows sufficient autonomy for smaller journeys, but is still not enough to be able to rival internal combustion motor cars over long distances[4].

The progress can still be made, however, on the level of specific energy, considering that, today, a Li-ion battery develops a specific energy equal to 140 Wh/kg, which is only 36% of the theoretic maximum energy (387 Wh/kg; Bruce *et al.* [BRU 12]).

This leaves a considerable margin for progress that important improvements in the future will make such a battery in electric vehicles more useful.

Increasing the insertion capacity of lithium in the electrode materials is therefore a major objective in research, and a number of teams are currently looking into this. The recent trivialization of the techniques involved in the production of graphene, as well as its remarkable conduction properties have made its use as a material for electrodes one of the more promising paths, that should soon yield important results.

6.1.2.1. *The functional originality of a Li-ion electrochemical cell*

An electrochemical cell is a mechanism that converts the energy from a chemical reaction into electrical energy, with a yield of nearly 100%, whereas thermal motors, depending on the system (petrol or diesel), have yields between 30 and 50%. This conversion takes place through two electrode reactions, each of them corresponding to the exchange of electrons between a chemical compound in a solution and a conducting surface that plays the role of an electrode.

This exchange of electrons happens for a particular potential that depends upon the energetic properties of the reaction. A more or less large difference in potential is thus established between the two electrodes: it is this difference in potential (electromotive force of the cell) that results in the production of an electric current (Box 6.1).

---

4 Nissan's electric car LEAF, released in 2012, has an autonomy of 160 km and a maximum speed of 144 km/h with a Li-ion battery that has a total energy of 24 kWh. If we take 140 Wh/kg as the Li-ion battery's effective specific energy, this means the minimal mass of material to be carried is 171 kg with approximately 1.8 kg of lithium, presuming that the discharge is continuous and equal to 3.5 V. For the sake of comparison, 1 kg of petrol contains 1,200 Wh/kg worth of usable energy, about 85 times that of the Li-ion system [WIN 04].

In the case of the Lithium-ion accumulator, the two electrodes, shown in Figure 6.3, are Li_xC_6 and $Li_{1-x}CoO_2$. Both the compounds, one carbon-based (graphite) and the other cobalt dioxide, present the interesting property of absorbing lithium ions in variable proportions (x between 0 and 1).

When immersed in an electrolyte (solution containing Li^+ ions), these two electrodes are characterized by very different potentials. Li_xC_6's potential is negative, whereas $Li_{1-x}CoO_2$'s potential is positive. The difference in potential between the two electrodes corresponds to the tension E supplied by the electrochemical cell.

As shown in Figure 6.3, a peculiarity of the electrochemical device is that the electric current is both electronic and ionic. In the external circuit, it is the electrons that are responsible for the current, whereas in the solution that separates the two electrodes, it is the positive lithium ions that travel in the opposite direction, and transport the charges that create the current.

During discharge, a difference in potential, E (electromotive force that can be used to run a motor), is obtained between the anode (electrode from which the electrons flow) and the cathode (electrode through which the electrons arrive). This tension remains more or less constant during most of the discharge, but falls rapidly when the levels of lithium in the carbon compound become very low.

The function of the battery, which corresponds to the discharge reaction, implies that the (–) pole be the anode and the (+) pole the cathode; the inverse is required during the charge sequence. A notable point is that between the charge and the discharge, the Li^+ ions come and go between the two electrodes, which has led to the system being referred to as a "rocking-chair battery", or "swing battery".

Currently, this system produces the highest voltages: 3.8 V between the two electrodes. Its specific energy, on the other hand, is a lot smaller than the specific energy obtained with $Li\text{-}O_2$ or Li-S batteries. For comparison, the theoretical energetic density per unit of mass of a Li-ion battery is 0.387 kWh/kg, that of a $Li\text{-}O_2$ battery with an aqueous and non-aqueous electrolyte is 3.583 and 3.505 kWh/kg, respectively, and that of a Li-S battery is 2.567 kWh/kg. It is its reliability that is above all its undisputed quality (Bruce *et al.* [BRU 12]).

6.1.2.2. *Nanotechnology to the rescue: the graphene solution?*

As we have just discussed, the functioning and autonomy of a Li-ion cell are directly linked to carbon's greater or smaller capacity to absorb lithium. The choice of graphite as lithium's host material is warranted as it is light and has a lamellar structure, which favors the insertion of lithium.

In the case of graphite compounds, the maximal amount of lithium incorporated cannot go above one atom of lithium for six carbon atoms, which is slightly less that 9%. This relies on all the layers of graphite being accessible to the lithium, which is not always the case, due to the strong cohesion between the sheets that make up graphite and slow the penetration of Li^+.

The concept of using carbon nanomaterials (NTs, graphene) instead of graphite seems to be wiser. They possess the same hexagonal lattice structure (six atoms of carbon) as graphite, but, when dispersed in a determined volume and for the same mass, they present a much larger specific surface and are therefore more likely to interact with a larger number of lithium ions.

Thus, while graphite's specific surface is only in the order of 10–20 m^2/g, single-walled carbon nanotubes (SWCNTs) reach specific surfaces of 1,315 m^2, double for graphene (2,600 m^2/g), which opens up the possibility of immense improvements in the insertion of lithium (Pumera [PUM 11]).

As a matter of fact this is a new branch of research that has itself developed a lot recently, first of all with CNTs and then, more recently, with graphene, for which the production methods are ever improving, and which are turning out to be cheaper to produce than CNTs for higher quantities produced.

Many innovations have been presented and evaluated meticulously. They show a considerable increase in the capacity for lithium insertion compared to the classic graphite-based system.

This accumulation of electricity, measured in milliampere-hours per gram of product (mAh/g), which is 350 mAh/g for graphite, can reach much higher values equal to 794–1054 mAh/g when using graphene-based composite materials. Such results imply a lithium composition close to LiC_2, or one lithium atom for two carbon atoms, instead of one for six in the case of graphite (Pan *et al.* [PAN 09])[5].

Other formulations have been envisaged, mixing powder-form graphene with silicon NPs, or metal oxides like cobalt, iron, titanium, tin, etc. (Sun *et al.* [SUN 11]).

This research is very recent. To date, they have not given rise to any commercial applications. International competition is very intense, and currently China is most involved in this domain, followed from a distance by the United States (Brousse *et al.* [BRO 12]).

Considering the increased research in the domain and the important progress made in graphene production techniques, it can be predicted that new Li-ion systems based on carbon nanomaterials might very soon make their appearance on the market.

6.1.3. *Electrochemical capacitors and supercapacitors*

As we have just seen, Li-ion batteries appear to be well adapted to providing high specific energies, which makes them all the more suited to use over long periods. However, considering the fact that the lithium insertion process is a relatively slow one, they are not able to release a lot of energy over a small time period. This means that they do not have the sufficient energetic capacity to start a powerful motor, for example.

Electrochemical capacitors (also known as electrolytic capacitors), which are very different systems from electrochemical batteries, present the peculiarity of functioning in a completely reversible and very rapid manner, allowing them to achieve extremely high levels of energy. Moreover, they are also capable of being charged and discharged practically infinitely, making them very reliable compared to all the other accumulator and battery systems.

---

5 1 mAh/g is equivalent to the quantity of electricity produced by 1 g of product, during 1 h and by an electric current of 1 mA. This is equivalent to 3.6 Coulomb/g, or a quantity of lithium consumed roughly equal to 0.3 mg. For example, the Sony NP BG1 battery used in mobile telephones weighs about 23 g, and delivers a tension of 3.6 V with a capacity of 950 mAh, which means the minimal amount of lithium contained in the carbon electrode is about 0.3 g.

The downside, unfortunately, is that their specific energy is low, and as a result they are poorly adapted to function over long periods of time. The diagram in Figure 6.4 compares the energetic properties of capacitors to accumulators.

It does show that capacitors can supply extremely high powers, but very low specific energies compared to accumulators and batteries.

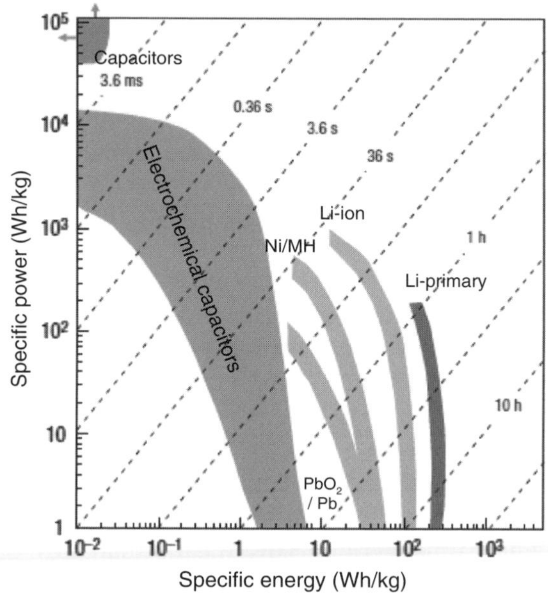

Figure 6.4. *Ragone plot. It establishes the correlation between the energy and the specific power of various electrical energy storage systems. The dotted diagonal lines show the length of the life of the battery of the capacitor for a specific energy given as a function of the required power. We can note that the capacitances (both classical and electrochemical) are characterized by very high specific powers, but function lengths corresponding to the discharge time are very short (adapted from Simon and Gogotsi [SIM 08])*

6.1.3.1. *Peculiarities of the electrochemical capacitor*

A traditional capacitor (not electrochemical) is a very simple electrical component comprising two metallic electrodes, separated one from the other by a thin layer of an isolating material called dielectric. It is a system that is capable, when a voltage is applied between the two electrodes, to store electrical charges and to return them fully when the electrodes are connected one to the other (Box 6.2).

Box 6.2. Capacitors

Dielectric capacitor: In its simplest form, it comprises two metallic plates (surface *S*) between which a voltage *V* is applied, and both plates are separated from each other by a thin piece of insulating material (dielectric).

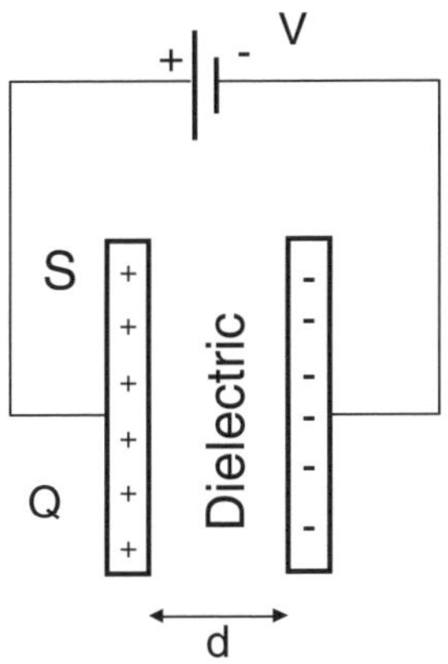

As a result of the applied tension *V*, an accumulation of opposite sign charges takes place on the two plates, and the current cannot pass from one plate to the other. The quantity of electricity (*Q*) stored in the capacitor is proportional to the applied potential and to a value called *C*, defined as being the capacitance of the capacitor.

This capacitance depends upon the geometrical characteristics of the capacitor and the nature of the material that separates the two plates. If the metallic plates are flat and parallel to each other, the capacitance *C* and the charge *Q* are proportional to the quantity *S*/*d*.

To this accumulation of charges there corresponds an energy that is itself proportional to the capacitance and to the square of the tension applied ($\frac{1}{2}CV^2$). When the two plates of the capacitor are connected to each other by an external conductor, the charges recombine extremely rapidly, and release the stored energy.

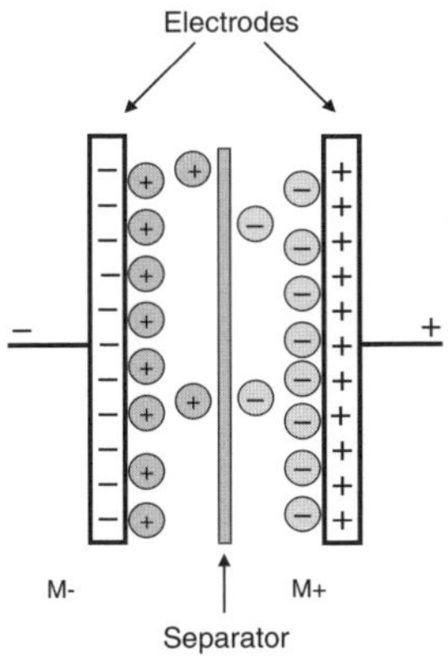

Electrolytic capacitor: The dielectric is replaced by a salt in a solution (electrolyte), dissociated into ions (+) and (–); the two plates are isolated one from the other by a plastic membrane (separator) permeable to ions. As a result of the applied tension, a fraction of ions (+) and (–) (cations and anions) are adsorbed, respectively, on the negative and positive plates of the capacitor. This is equivalent to two capacitors in series, of capacity C_1 and C_2, resulting in two metal/ion interfaces, *M*–/(+) and *M*+/(–). The distance separating the ions from the different plates is largely inferior to that found in a dielectric capacitor, which explains the high values of capacitance obtained with electrolytic capacitors.

The electrolytic capacitor, despite being based on the same principles, differs from the dielectric capacitor by the presence of an electrolyte (liquid or gel containing a salt dissociated into ions (+) and (–), which correspond, respectively, to the cations and the anions) that is in contact with the electrodes and that replaces the dielectric.

In the absence of a potential difference between the capacitor's electrodes, the distribution of the ions in the solution is homogeneous, and the electrical neutrality is maintained at every point of the solution.

This is no longer the case when a tension is applied between the capacitor's electrodes. The conductors become, respectively, negatively and positively charged, in the same way as a dielectric capacitor, but as a result of this the ions start to become asymmetrically distributed in the areas immediately surrounding each electrode/solution interface (Figure 6.5).

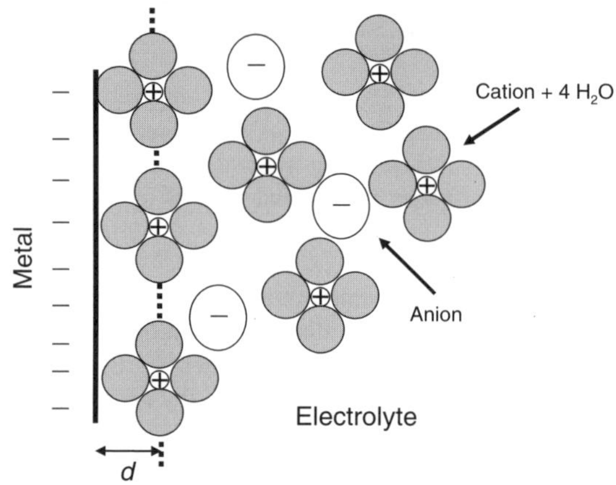

Figure 6.5. *Simplified image of the interface between a negatively polarized metal and a solution containing a salt, made of positive (cations, +) and negative (anion, –) ions. The metal's surface, given a negative potential, is negatively charged. The solution's (+) ions (surrounded by four water molecules) are attracted by the electrodes, negative charges, while the negative ions (usually larger than the cations and not surrounded by molecules) are pushed toward the inner part of the solution. The cations position themselves in a plane (drawn in dots) that is at a distance d (a few angstroms) from the surface. The distance of separation is determined by the dimensions of the water molecules that surround the positive ion; the combined positive (solution) and negative (metal) charges make up the equivalent of a very high capacitance capacitor, due to the very low value of d*

In this way, a negatively charged electrode induces an electric field that attracts ions of the opposite sign (cations) and pushes away the negative ions (anions). The cations therefore move toward the surface of the electrode, although they cannot get any closer than a distance *d*.

This distance *d*, which is well-known to electrochemists, corresponds to the plane in which the ions are as close to the electrode as possible (also known as the outer Helmholtz Plane[6]). The separation of the charges that appear between the electrode and the solution constitutes the double electrochemical layer, and can be considered as the equivalent of a capacitor.

In practice, an electrochemical capacitor is created by putting together two electrodes (metallic sheets) and separating them by a thin, porous, insulating membrane (separator) imbibed with electrolytes. This is equivalent to putting two capacitors in series, each consisting of a metal–solution interface.

The very meager distance separating the charges at the metal–solution interface (a fraction of a nanometer) gives the system a very high electrical capacitance (*C*), and therefore the possibility of storing a lot more energy compared to the classical capacitors made of an insulating dielectric, the minimal thickness of which is at the best in the order of 10 nm.

This has the effect of greatly increasing the amount of charge that can potentially be stored, and, consequently the amount of energy that can be stored, which is much more than with non-electrolytic capacitors. For this reason, they are often referred to as supercapacitors or ultracapacitors.

6.1.3.2. *The developments and the state of the art*

The amount of energy stored in commercial supercapacitors (about 5 Wh/kg) is a much less than the energy stored in commercial Li-ion batteries (between 100 and 150 Wh/kg). On the other hand, due to the absence of a chemical reaction, they are able to deliver extremely high levels of power in small amounts of time (between 1,000 and 5,000 W/kg), whereas the Li-ion batteries never go over more than 150 W/kg.

In a few cases, powers of 100,000 Watts/kg can be obtained, but with much lower energy densities, around 0.1 Wh/kg.

As explained above, these supercapacitors are used primarily as energy sources for systems that require a high level of instantaneous power to get started. They can also be put through a large number of charge–discharge cycles without being damaged (often more than 100,000 cycles, which is a

6 H.L. von Helmholtz (1821–1894) was a German physicist remembered primarily for his work in the areas of thermodynamics, physiology and acoustics.

hundred times more than electrochemical batteries), and for this reason are very reliable in the long run.

As the storage capabilities of these capacitors is their main weakness, all research efforts are concentrated on improving this property, and the simplest way to increase an electrolytic capacitor's capacitance is to artificially increase its specific area. The use of "porous" electrodes is a first solution.

6.1.3.2.1. Porous electrodes

A porous electrode is easily obtained by depositing on the surface of the electrode "active" conducting coal particles of submicrometric dimensions and bathing them in the electrolyte. This considerably increases the specific surface of the electrode, which theoretically becomes equivalent to the combined surfaces of all the carbon particles, which are very good conductors.

The capacitances achieved are measured in Farads per gram (F/g), a unit that links the accumulated electric charge (Q) to the potential (V) ($Q = CV$), and to the amount of energy stored $E = ½CV^2$.

A recent idea is to replace the carbon particles with graphene. Its specific surface is very high at 2,630 m^2/g, and its electrical conductivity is much greater than the "active coals". This might allow very strong capacitances per unit of weight, theoretically going as high as 530 F/g (Brousse [BRO 12]).

In practice, the achievement of such a value is still a long way off. Depending on the source and on the technique used in preparing the graphene, the specific capacitances in an aqueous medium are between 120 and 270 F/g. For a tension of 1 V, this equates to a specific energy of approximately 20–40 Wh/kg, still very far from the specific energies of the Li-ion batteries.

Other improvements might be made by making hybrid supercapacitors, known more technically as "pseudo-capacitors". The principle is to simultaneously combine two effects: one is purely related to capacitance and the other is related to an electrochemical faradic process (transfer of electrons at the metal–solution interface). The use of CNTs and graphene was recently tested, and appears to be well suited for this kind of device.

6.1.3.2.2. Graphene or CNTs for the "pseudo-capacitors"?

Recent work made use of the latest techniques in making graphene from its oxidized form, combined with the adsorption by sheets of graphene of metal oxide NPs (MnO_2) obtained through a treatment with microwaves of permanganate ions (MnO_4^-). The results were promising, largely surpassing the performances of graphene-only supercapacitors (Figure 6.6).

Particles of manganese oxide (MnO_2), adsorbed by sheets of graphene, can in this way be reduced (acquire an electron) and re-oxidized (loss of the acquired electron), and can therefore exist in the form of a negatively charged particle, or in their neutral state.

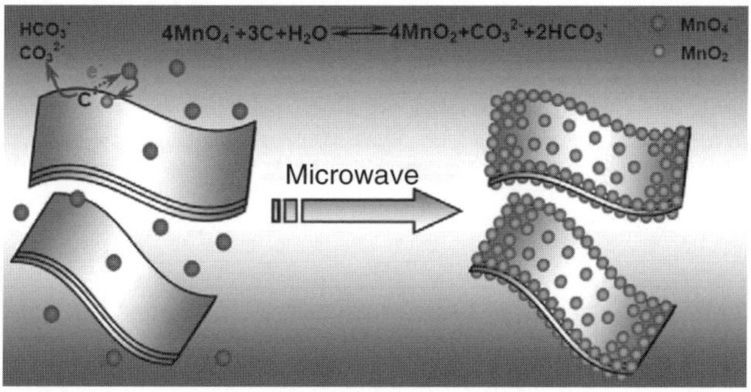

Figure 6.6. *Process of "decoration" of the sheets of graphene by manganese oxide (MnO_2) nanoparticles (adapted from Yan et al. [YAN 10]). For a color version of this figure see www.iste.co.uk/lacaze/nano.zip*

When they acquire a negative charge, the cations (H^+, Li^+ or Na^+) are attracted to the NPs, and are distributed over most of the surface area of the graphene. However, when the NP returns to its state of electrical neutrality (a process accompanied by an exchange of electrons, see Figure 6.7), the cations are leaving the graphene's surface.

The advantage of this process is that it is a lot faster than a classical insertion reaction. The system's capacitance is considerably increased, and the speed of charge and discharge, although still slower than in the case of an electrolytic capacitor, is nevertheless faster than for a Li-ion-type insertion battery.

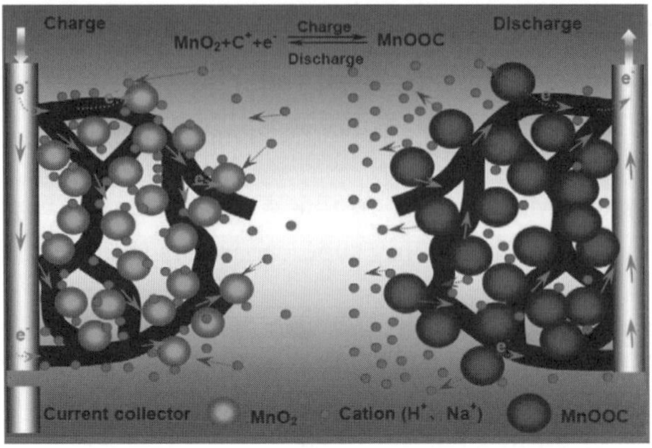

Figure 6.7. *Charge and discharge mechanism of the pseudo-capacitor made of sheets of graphene and NPs of MnO_2. During the charge process (left electrode), the MnO_2 NPs are negatively charged and the cation C+ come to anchor themselves on the NPs, forming a neutral compound $MnO_2(-)C(+)$, written MnOOC. At the moment of discharge (right electrode), the inverse reaction takes place, with MnO_2 returning to neutrality, and the cation C+ being expulsed (adapted from Yan et al. [YAN 10]). For a color version of this figure see www.iste.co.uk/lacaze/nano.zip*

Similar systems have also been presented recently by Japanese teams, using nanofibers of CNTs on which $Li_4Ti_5O_{12}$ nanocrystals (result of the insertion of lithium into titanium oxide) are deposited. According to the authors, these "nanohybrid" supercapacitors perform much better than the classical systems (Naoi *et al.* [NAO 12]).

Specific capacitances of over 1,000 F/g have been announced, although it may be wise not to get too enthusiastic and to wait until practical applications make their way onto the market. Therefore, the multitude of innovative works focused on the development of new nanomaterials should lead to much improved capacitor-style devices in the near future.

6.1.4. *Conclusions*

The electrochemical storage of energy has for a long time been a research area of primary importance. The development of renewable energies, the multiplication of mobile electronic systems of all types, and more recently the gradual replacement of thermal motor vehicles by fully electric vehicles all undeniably increase the demand for a range of varied devices, and stimulate industrial research toward systems that can provide ever-increasing energy storage. This research, which is tightly linked to material science, appears to benefit from progress made in the production of all sorts of nanomaterials, and the appearance of important applications can be expected in the next few years that should considerably improve the performance of storage systems.

6.2. The conversion of solar energy into electrical energy

This is another research area, just as important as the previous one, that is developing under the increasing ecological "pressure" and also due to the fact, that new energetic solutions must be found to replace unavoidable exhaustion of fossil fuels.

An obvious inexhaustible source of energy is the Sun. In 1 s, it sends to Earth the considerable amount of roughly 120,000 TW[7], whereas our average consumption is approximately 13 TW. This should be enough to reassure us of our energetic future, as only 0.01% of this energy would need to be captured to avoid any kind of shortage.

In reality, we are nowhere near this. In 2011, on a global level, solar energy produced as electrical energy (photovoltaic) only represented 34 GW (0.034 TW), which is a little less than 0.3% of the global energy power consumption. This corresponds to a minute fraction of the Sun's energy power received, being only 3/10,000 (Philibert [PHI 05], Lewis and Nocera [LEW 06]).

This is very little, and can be explained by the fact that only a small number of countries currently have conditions that favor solar energy in place.

However, the overall trend is a rise in solar energy (between 2009 and 2011, there was an increase in photovoltaic energy of 70%). Moreover, rising prices of fossil fuels, as well as the active development of research in the domain, promising low-cost devices with increasingly high yields, should lead to a rapid growth in this field, as suggested by several international organizations (2011 report by the International Energy Agency [IEA 11], 2012 report by the European Photovoltaic Industry Association [EPI 12]).

---

7 A terawatt (TW) is equal to a power of a thousand billion watts (10^{12} W), the watt being the unit of power, equal to 1 J/s.

Different technologies are used in the conversion of solar energy: "passive" technologies that involve heating absorber materials to retrieve the heat and "active" technologies that involve transforming solar energy into electricity.

It is obviously this active technology that is the most interesting, and it is also in this domain that the most important progress is expected.

Several branches are undergoing considerable development: the use of silicon, mostly commercialized already, but still relatively expensive, and the use of organic compounds, which could potentially be used in large-scale development and at a low cost, but which suffers from a conversion yield too low to be competitive with the use of silicon.

Nevertheless, it is in this domain that progress must be made in the future, and the use of nanomaterials has already contributed a number of important improvements.

6.2.1. *The principle of the conversion*

A photovoltaic device designed to transform light energy into an electric current requires first the successive completion of two actions:

1) a process of photoionization that leads by irradiation of a material to the creation of an electron-hole pairing (electron-positive charge) called an exciton;

2) a process of separation and transport of the charges thus produced, which are then, respectively, collected toward a positive and a negative electrode and produce an electric current.

6.2.1.1. *The photoelectric effect and its history*

A metallic surface placed in a vacuum and irradiated by light emits electrons as soon as the wavelength reaches a value below a certain threshold characteristic of the metal. The effect, known a long time ago, occurs with metallic materials, and Einstein described it in 1905, for which he received the Nobel Prize in 1921.

The principle relies on the fact that light must not only be considered as a periodic wave that oscillates at a frequency of ν but also as a carrier of energy particles, photons, that have an energy measured in quantas (hv) directly proportional to the universal constant h (the Planck constant) and the radiation vibration frequency v.

When the photons that accompany the light rays collide with atoms of the material, they release their energy to the electrons linked to the atoms of the surface of the material (Figure 6.8). If this energy gained by the electrons is intense enough, they can extract themselves from the metal, in which case the emission of electrons into the vacuum takes place.

This emission can take place only for a minimal value of energy of the photons corresponding to an energy threshold, characteristic of the metal, and which corresponds to the physical work of the electrons done in leaving the metal.

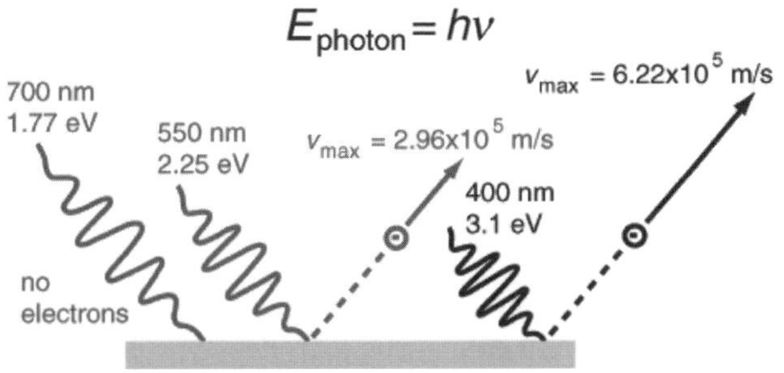

Figure 6.8. *Photoelectric effect. By irradiating the potassium (K) surface in a vacuum with radiations of different wavelengths corresponding to regions of the infrared (700 nm), visible (550 nm, the color green) and UV (400 nm) spectra, the emission of electrons only takes place upon reaching the threshold around 2 eV, which corresponds to the wavelength 618 nm (red color). The infrared radiation with a wavelength of 700 nm does not have the sufficient amount of energy (1.77 eV) to extract the electrons from the metal. The speeds* v_{max} *transferred to the electrons emitted are higher for higher energy values of the photons (the square of the speed is proportional to the energy difference* $hv - eV$*)*

It is this unique energetic trait of the photoelectric effect that for a long time intrigued physicists, and was finally solved by Einstein; it also provided evidence supporting Planck's quantum theory[8].

8 Max Planck (1858–1947) was a German physicist, winner of the Nobel Prize in 1918, who made the postulate on quanta that establishes the discrete (not continuous) nature of energy.

6.2.1.2. *Photoionization of a semiconductor and collection of the charges at the electrodes*

This double effect takes place when a transparent metal electrode (*M*) is associated with an SC, or even when two SCs of different nature n and p (n-SC/p-SC) are associated together, forming what are, respectively, called a Schottky diode and a p–n junction[9].

Adams and Day were the first to observe such an effect in 1877 with selenium. A little later, Fritts created in 1883 the first photovoltaic solar panel by illuminating a selenium crystal covered in a thin layer of gold. This yield was of roughly 1 to 2%[10].

At this point, the question that springs to mind is how does light behave when it reacts with an SC?

The behavioral principles of the photons remain the same as described previously in the case of the photoelectric effect. The electrons of the SC are distributed into different bands of energy and the energy transferred from the photons to the electrons is used to induce transitions between the different energy levels of the SC.

Let us remind ourselves that an SC consists of two bands of energy (a VB, totally occupied by electrons, and a CB, nearly completely unoccupied) separated by a band of forbidden energy with a width of E_g (Figure 4.1).

By carefully choosing the energy of the photons, slightly greater than E_g, the photoelectric effect is basically an "excitation" of the electrons of the VB toward the conduction band, which consequently creates an electron-hole pair, which means a positive charge is in the VB, and an electron is in the CB (Figure 6.9).

---

9 An SC n-Si or p-Si is a silicon SC, doped respectively by atoms that either have an extra electron (phosphorus, arsenic) or an electron less (boron) than the Si atom. With an n-Si type SC, it is the electrons (negative charges) that are primarily responsible for conductivity, whereas on the other hand with a p-doped SC (p-Si) conductivity is due to the presence of "holes" (positive charges).

10 The first discovery of the photovoltaic effect is even older and is attributed to E. Becquerel in 1839. The effect was, however, observed under different conditions and in the case of a liquid electrolysis cell under illumination.

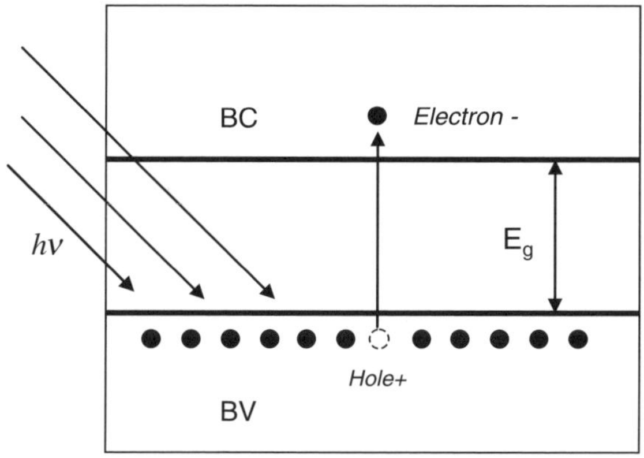

Figure 6.9. *Illustrated principle of the formation of an electron-hole pair in an SC with E_g the width of the forbidden band under the effect of radiation of energy hv greater than E_g*

Thus, in the case of silicon, for which E_g is roughly 1.1 eV, a large part of visible light with wavelengths between 400 and 700 nm possesses enough energy to cause this energy transition of the electrons from the BV toward the BC.

Without any additional effect, the recombination of the hole and the electron is inevitable, and corresponds to the return of the electron to the VB and the annihilation of the positive charge, which consequently causes the absence of any electrical current and instead only the release of heat.

In fact, by making a metal-SC or n-SC/p-SC junction, a potential barrier is spontaneously created at the interface of the two materials, and it is this potential difference established between the two electrodes that is responsible for the separation of the charges, by attracting the positive charges (holes) toward the negative electrode and the negative charges (electrons) toward the positive electrode (Figure 6.10).

The result is the appearance of a photocurrent and of a photopotential (usually between 0.5 and 1.6 V) characteristic of the assembled materials. This is the photovoltaic effect characterized by a yield of energetic conversion equal to the ratio of the power of the electrical energy produced and the power of the radiation received.

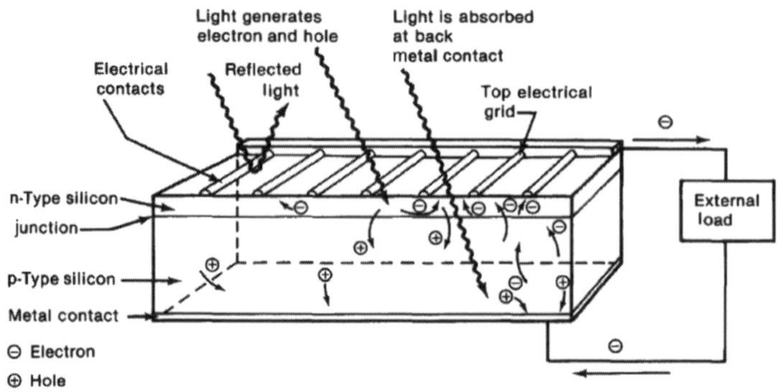

Figure 6.10. *Illustrated principle of the photovoltaic effect of a p–n junction. The irradiation of the SCs takes place through a metal grid that plays the role of the positive electrode, but a transparent conducting ITO-type (indium-tin oxide alloy) electrode can also be used. The bottom negative electrode is a layer of metal deposited on the p-type Si (adapted from Hersch and Zweibel [HER 82])*

6.2.2. *The inorganic route based on mineral semiconductors*

Using a simple montage involving a single silicon p–n junction, the width of the forbidden band being 1.1 eV, the theoretical yield of the conversion cannot be greater than 34%. In practice, and despite there having been constant progress over the years, the conversion rates are much lower. They depend upon the quality of the silicon. In 2010, the yields were at the most 25% (monocrystalline Si), 20% (polycrystalline Si) and 12.5% (amorphous Si) (National Renewable Energy Laboratory (NREL) report, Best Research-Cell Efficiencies, 2010, USA).

The main commercial devices are silicon and p–n junction based, and the silicon used is usually amorphous silicon, which is much cheaper to produce than monocrystalline or polycrystalline silicon, but is worse performance-wise.

Multijunction (MJ) systems have been designed with the objective of achieving yields above the theoretical limit of 34%, which applies only to single junctions. This involves converting all the IR range of the solar spectrum that constitutes an important part of the Sun's energy into electrical energy, a feature which is not achieved with the silicon-based single junctions (Figure 6.11).

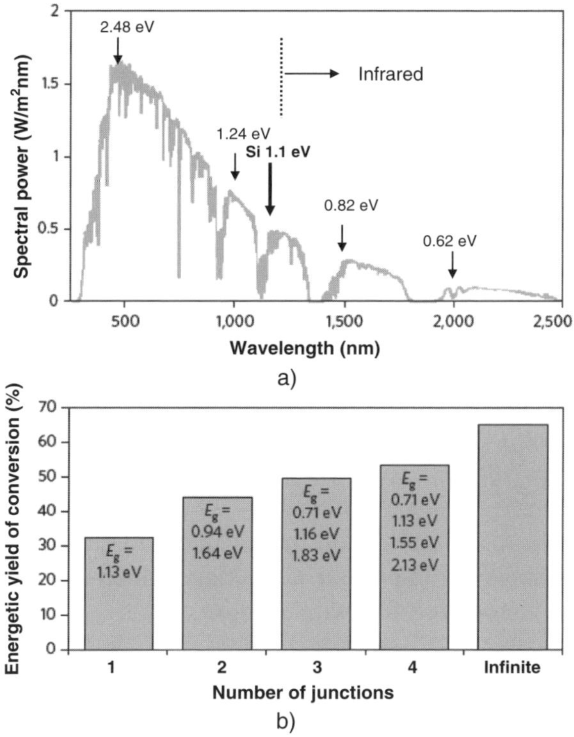

Figure 6.11. *a) Solar spectral power as a function of the wavelength (note that the powers on the x axis are divided by the corresponding wavelength, which artificially understates the energies relative to the infrared region of the spectrum, located above 1,250 nm). The arrows distributed over the radiation curve point to the corresponding energies of the wavelengths and show that a large part of the infrared spectrum (λ > 1,250 nm or hv < 1eV) cannot be collected with a silicon-based junction with an E_g equal to 1.1 eV; b) energy conversion yields as a function of the number of junctions (between one and infinity) with a layering of the values of E_g as a function of the number of junctions (adapted from Sargent [SAR 09])*

This is obtained by putting in series several cells comprising SCs with smaller forbidden bands. However, the setups are tricky and very expensive, and were mainly designed for certain applications, in particular, such as the electrical equipment in satellites.

The conversion yields are then very high, and can still be increased by concentrating the light using optic devices beforehand (NREL 2010 report). Under these conditions, values much higher than the theoretical limit (64%) for a single junction can be reached. Such systems use a very complicated technology, consisting to deposit multiple thin film layers of different SCs, comparable to the technology used in the industrial production of thin film transistors. For the most part they must be considered as laboratory-scale productions.

Commercially, such systems prove to be very costly; less effective but less expensive routes tend to be preferred. Recently, a detailed description of the many systems used in inorganic photovoltaic technologies was compiled by Lincot [LIN 11].

6.2.3. *The organic route*

Until recently, the dominant technology in solar cells was silicon based. It is a well-developed technology, but expensive, that requires the transformation of silica into silicon, a reaction that uses a lot of energy.

The concept of replacing silicon by cheaper and easily available organic materials therefore is an important challenge, which has already been the subject of intensive research for two decades.

Several devices are already under study. Foremost they are fully organic solid systems and mixed inorganic-organic systems of electrochemical nature, known as Grätzel cells.

Fully organic systems have another advantage over inorganic silicon systems: they can be made on flexible plastic surfaces, which further widens the list of their possible applications. Conversion yields of around 6–7% are still too small to warrant their use commercially. Nonetheless, the intensive research currently underway could lead to their commercialization in the near future.

The second option, based on Grätzel cells, is already reaching yields of roughly 12%, allowing it to compete with silicon based technologies. This has led to commercial development; however, use is technically difficult, as it requires an electrochemical step involved in the energetic conversion of light.

6.2.3.1. *Organic photovoltaic cells*

The principle is the same as for the silicon-based SCs. The photovoltaic effect relies in this case on the use of organic SCs, which unfortunately have very inferior electrical properties to silicon.

Although the initial production of an electron-hole pair is a relatively easy process, the definite separation of the charges and their progression toward electrodes of opposite signs to them is actually less obvious, and is met with a number of "hazards", which ultimately results in smaller conversion yields.

The decrease in the conversion yields is due for a large part to the fact that the mobility (speed of a charge under the effect of an electrical field) of the charges created by the light is usually a lot smaller than in the case of silicon or all other inorganic SCs.

As a result, the parasitic phenomenon of charge recombination is more important, causing a decrease in the yields of free electric charge, and therefore a decrease in the photocurrent.

The search for organic materials with high charge-carrier mobilities is very important in determining the efficiencies of such devices.

Currently, the most effective photovoltaic systems consist of a mix of two molecular materials that differ in their electrical properties: one is an electron donor (D) and the other is an electron acceptor (A).

D is usually a conjugated polymer (also called an electronic conducting polymer, Box 6.3) that has the property of strongly absorbing visible light. Its characterizing feature is being formed of carbon chains made of alternating double and single bonds, such as poly-para-phenylene-vinylidene (PPV), very common in this type of cell.

The conjugation of the single and double bonds causes the electrons to become very mobile and more excitable in the presence of light.

Like for the SCs, under the effect of light stimulation, the electrons follow a transition from a low-energy level (stationary state) toward a higher-energy level (excited state). If the acceptor (A) is in the proximity of D, and if it possesses an empty energy level favorably placed in relation to D's excited state, it can receive the electron and transform into A^- at the same time that D becomes D^+.

Box 6.3. Conducting polymers

Polymers make up a class of compounds with very high molecular masses, due to the near-infinite repetition of a same molecular motif (which itself has a small mass). The concept, which only appeared late in the history of chemistry (1920), was discovered by the German chemist Hermann Staudinger (Nobel Prize 1953). Nowadays, they can be found nearly everywhere in the form of plastics (used in packaging for example). This is mostly polyethene $(–CH2–CH2–)_n$, polyvinyl chloride $(–CH2–CHCl–)_n$, polycarbonates, polyamides, etc. They differ from each other in their mechanical properties, their thermal stability domains and their resistance to different solvents, but they all have an electrical property in common, that is they are excellent insulators.

Conjugated conducting polymers are more recent and were discovered in the 1970s by MacDiarmid, Heeger and Shirakawa, all three of whom were awarded the Nobel Prize in 2000. By using polyacetylene $(=CH–CH=CH–CH=)_n$, a conjugated polymer characterized by the repeat of a carbon chain made of double and single bonds, they showed that it was possible to modulate its conductivity over an extremely large range, going from insulators to metallic conductors.

This exceptional property is the result of a doping phenomenon, caused by the addition of an oxidizing reagent such as iodine. By oxidizing the polyacetylene's chain, "holes" (positive charges) are created. These holes can move along the chain very rapidly, and they are the holes that give the polymer its high electric conductivity.

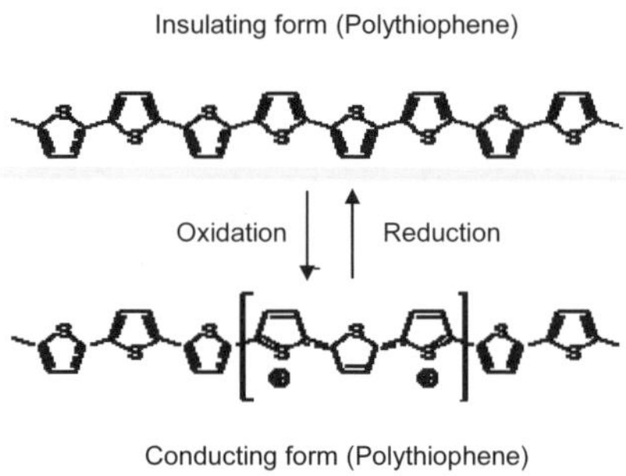

Since this discovery (1977), the same phenomenon has been observed in other conducting polymer families such as polyparaphenylene (PPP), polypyrrole (PPy), polyaniline (PANI), polythiophene (PT), poly-2,4-ethylene-dioxy-thiophene (PEDOT).

After the charges have been separated, they must be moved quickly toward the electrodes. Hence, there is a need to have on hand both electron donor materials that are good conductors of positive charges (holes) and electron acceptor materials that are good electron conductors. Polymer conductors (similar to p-type SCs) and the fullerenes (comparable to n-type SCs) are the very materials that best fulfill these functions, which is why their conjoint use brings a considerable increase in the conversion yield of these devices (Figure 6.12).

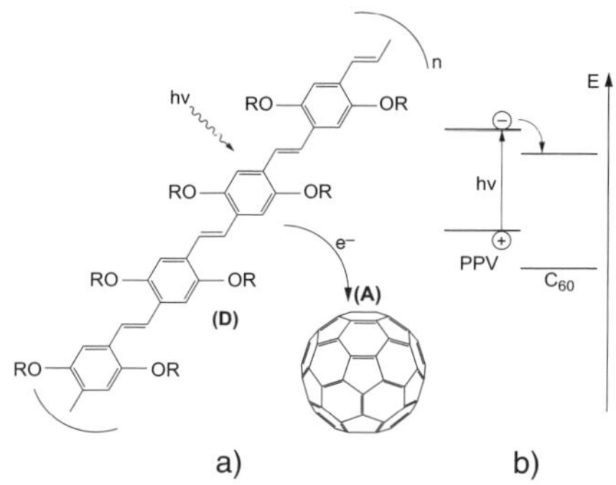

Figure 6.12. *a) Polymer derived from the electron donor (D) PPV, and the electron acceptor (A) fullerene (C_{60}); b) energy diagram showing the transfer of charges from D toward A after the light stimulation of D (adapted from Günes et al. [GÜN 07])*

From a practical point of view, it must be noted that the process of separation and the transport of charges can take place significantly only if the separation distance between donors and acceptors is optimal, which implies that the two networks, donor and acceptor, are strongly interpenetrating each other. This is a complex problem, far from solved, for which there is currently no definite answer.

Nonetheless, due to its simplicity and the low cost of its production this type of cell is very interesting. The goal is to elaborate photosensitive layers using coating techniques, equivalent to those that are used for applying paint to the surface of a material.

Figure 6.13 represents a typical example of "plastic" photovoltaic cell made by mixing poly-3-hexyl-thiophene (P3HT) with the most commonly used acceptor, a derivative of fullerene C_{60}, PCBM.

The intimate mixture between P3HT and the fullerene derivative (PCBM) is fundamental and largely determines the qualities of the junction, which is then called a bulk heterojunction (BHJ). The difficulty of setting up a perfect interpenetration of the two networks, which requires rigorous experimental protocol, in which a number of physicochemical parameters are involved, is partially the reason for which varied yields have been reported by different teams.

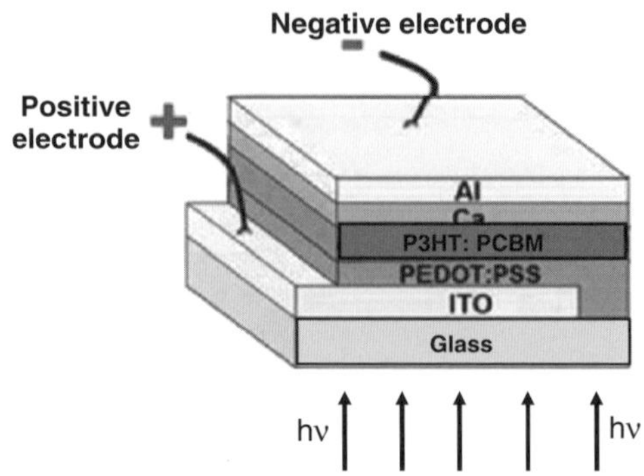

Figure 6.13. *Conceptual diagram of an organic solar cell made of a photosensitive layer P3HT: PCBM a few microns thick. The electrical contact between the sensitive layer and the ITO is improved by a thin layer of another conducting polymer with a similar structure (oxidized PEDOT, transparent, associated with polystyrene sulfonate – PSS). A thin layer of calcium (Ca) is also deposited on the photosensitive layer so as to improve the electrical contact with the positive aluminum electrode (adapted from Li [Li 12])*

The conversion yields are still modest (maximal values were announced to be around 7–8%), and although the progress is slow, it is continuous. Between 1995 and 2011, the yields have indeed progressed from 1 to nearly 8%, allowing hoped for yields over 10% in the years to come, which would make this type of solar cell very competitive in terms of commercial exploitation [LI 12].

6.2.3.2. *Grätzel dye-sensitized solar cells (DSSC)*

This is another branch of research, directly inspired by the process of photosynthesis, presented by Grätzel in the 1990s.

The concept involves using an n-type SC, of low cost and abundant, like titanium dioxide (TiO_2)[11] with which a photovoltaic effect might be achieved like with silicon.

Unfortunately, the problem is that this SC is characterized by a large forbidden band (about 3 eV, or three times more than silicon), which excludes directly separating the charges using visible light.

---

11 Titanium dioxide is a non-toxic industrial product, very popular in the paints industry as a white pigment. It is also present in sunscreens due its UV-ray absorption properties, in autocleaning and anti-dirt glass. The absorption of UV by TiO_2 causes photochemical reactions that destroy polluting molecules and absorb hydrophilic groups at the surface of the glass, making it hydrophilic itself and self-cleaning.

However, this inconvenience can be minimized by "sensitizing" the SC with a dye, so as to make effective the separation of the charges by a "cascade" effect. This is done by adsorbing at the surface of the SC a colorant characterized by a property of strongly absorbing visible light.

The conversion setup does, however, require that the dye returns to its initial state immediately after becoming excited by the light. This requires the presence of a mediator (redox system), that by an electrochemical transfer of electrons regenerates the dye to its initial state. This mediator is itself regenerated at a platinum counter electrode, which completes the electric circuit and induces the photocurrent (Figure 6.14).

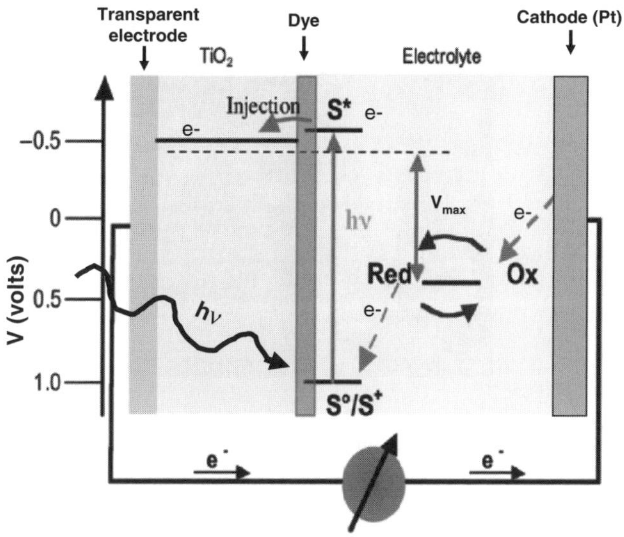

Figure 6.14. *Principle behind a Grätzel colorant cell. Excitation by light (hv) of the dye S° induces the formation of an excited state S*, which corresponds to an electron jumping from the fundamental state S° toward the excited level S*. The electron migrates very rapidly through SiO_2's conduction band, and then moves toward the transparent electrode. During this electronic transfer, S° loses an electron and transforms into S^+. The system can only continue to function if S^+ returns to its initial state S°. This transformation is caused by the redox couple (Red/Ox) that plays the role of reduction-oxidation mediator. In the present example, this is the couple iodide/iodine (I^-/I_2). In this case, the mediator transforms S^+ into S° through the redox reaction $I^- + S^+ \rightarrow S° + ½ I_2$. I^- (Red) is itself regenerated to the cathode by an electron transfer: ½ I_2 (Ox) + $e^- \rightarrow I^-$ (Red). The energy absorbed by the colorant is thus partially returned in the form of a photocurrent and of a photopotential V determined by the SC and the redox system (adapted from Grätzel [GRÄ 09])*

This idea was developed by Grätzel at the Ecole Polytechnique Fédérale at Lausanne, and has since led to the high performances photovoltaic conversion devices seen today, and to considerable commercial use, especially in Japan.

It must be noted that these high conversion yields were not obtained immediately. It was only after the initial film of titanium dioxide was replaced by assembly of TiO_2 nanospheres (between 15 and 20 nm in diameter) that the photoconversion yields greatly improved, going rapidly from a few percent to over 10%, the principal reason being the considerable increase in the quantity of dye adsorbed at the surface of the SC (Figure 6.15).

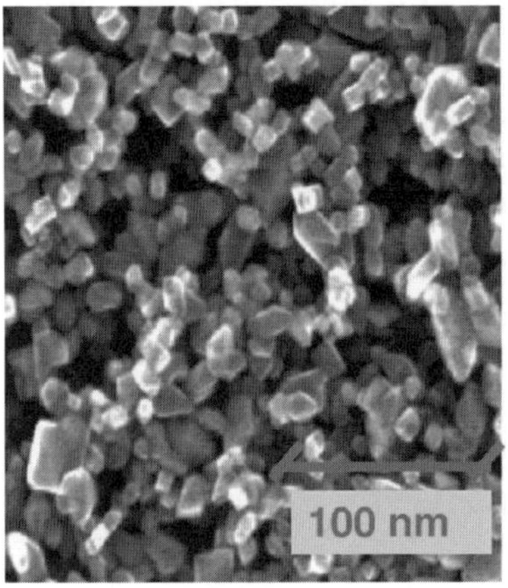

Figure 6.15. *SEM image of a transparent mesoporous layer a few tens of microns thick, made of nanoparticles of TiO_2, roughly 20 nm in diameter, on which the dye is adsorbed (adapted from Grätzel [GRÄ 09])*

Many improvements have been progressively brought to the device, as much in the choice of dyes as in the nanostructuration of the titanium dioxide layer.

Huge progress has been made in this way by replacing the all-organic dyes with ruthenium complexes (relatively expensive), which offer increased chemical stability and also absorb an important fraction of the infrared spectrum, which helps to increase the energy conversion yields.

Concerning the mesoporous layer of titanium dioxide, several nanostructuration variations have been tested with the goal of improving the intensity of the photocurrent. Mesoporous layers made up of NTs, nanorods or even hybrid "core–shell" NPs with a TiO_2 core and a shell made of another material have been evaluated, the objective being to decrease the occurrence of recombination of the photogenerated electrons (Zhang and Cao [ZHA 11]).

Moreover, the introduction of large (200–400 nm) particles of TiO_2 into the center of the layer has resulted in a significant increase in the diffusion of the light and, consequently, the intensity of the photocurrent.

All these improvements made in the last 20 years have resulted in a very reliable photovoltaic system, with photoconversion yields satisfyingly reproducible (between 8% and 10%), as long as the preparation protocol is rigorously observed. Function over longer periods has proved satisfactory and the cost should be less than for silicon concerning possible large-scale commercial exploitation. The possibility of making panels of huge dimensions, using techniques derived from the paints industry, could prove to be the advantage over silicon-based technologies.

6.3. Fuel cells

A fuel cell is a particular type of electrochemical cell, designed to continuously produce electricity as long as it is permanently supplied with a fuel and an oxidizing agent, in a situation comparable to a thermal motor.

Just like a classical electrochemical battery, a fuel cell requires two different electrode reactions: (1) the oxidation at an anode of a reducing compound called fuel (dihydrogen, alcohol, hydrocarbon, etc.) and (2) the reduction at a cathode of an oxidant (dioxygen or air). To avoid the mixing of gases within the cell, the anode and the cathode are separated by an electrical insulating porous membrane that is permeable to the electrolyte. The other difference with an accumulator comes down to the fact that reagents to be transformed are stored outside and are continuously injected into the cell.

Essentially, two categories of fuel cell can be distinguished depending on the operating temperature.

Those that use a liquid electrolyte operate at the temperatures between –20 and 100–150 °C, namely the H_2/O_2 proton exchange membrane fuel cells (PEMFCs), direct methanol fuel cells (DMFCs) in which the fuel is methanol and alkaline fuel cells (AFCs), which exchange hydroxyl ions (OH^-).

Those that operate at the high temperatures between 600 °C and 1,000 °C, namely molten carbonate fuel cells (MCFCs) and solid oxide fuel cells (SOFCs). In both cases, the separating membrane is an oxide that becomes permeable to ions at very high temperatures[12].

With dihydrogen as the fuel, the final product of the electrochemical reaction is water, making the systems completely pollution-free, which is a significant advantage for applications in fully electric vehicles and in CO_2-free energy production. This is not the case with methanol fuel cells that operate at temperatures between 20 °C and 60 °C, and release carbon dioxide (CO_2) and water as final products of the reaction.

Nevertheless, these two different systems have both advantages and disadvantages.

In the case of MCFCs or SOFCs, the high operating temperatures considerably accelerate the speed of transformation, which gets rid of the need for expensive catalysts. This is not the case with cells operating at ambient temperatures like PEMFCs, which require the electrodes to be coated in a precious metal such as platinum. On the other hand, the high temperatures to which the materials are exposed make them particularly vulnerable to corrosion. As well as this, their lack of instantaneity in beginning the reaction make any eventual application in portable electronic devices or vehicles impossible.

The PEFMC systems that function at a whole range of temperatures between –20 and a hundred or so degrees are clearly better adapted to supplying power to portable systems than high temperature cells.

Their usage in electrical vehicles is already well established, and many automobile manufacturers offer fully electric electrical fuel cell vehicles (EFCVs). The performances exceed by far those of the accumulators, both in their large energy storage, allowing good functional autonomy, and in their short recharge times, comparable to the times needed to recharge the fuel in a thermal motor vehicle.

---

12 In MCFCs, ionic conduction is guaranteed by carbonate ions (CO_3^{2-}), whereas in the oxide cells, which function at temperatures 800 °C and more, the O^{2-} ions are the ions responsible for the ionic current. Typically, their functional characteristics are 150 mA/cm^2 and 0.8 V at 600 °C for the former, 1 A/cm^2 and 0.7 V at 800–1,000 °C for the latter [WIN 04].

Such systems have already been tested by the public transport services in several countries. Since 2010, the San Francisco region has been equipped with a network of fully electric buses that use fuel cells with a power of 120 kW (UTC Power) and 21 kWh (Ener 1) lithium accumulators. Furthermore, hydrogen terminals installed around the city allow vehicles to fill up on fuel extremely quickly, ensuring the vehicles run continuously. Finally, to complete the ecological nature of the project, the stations are supplied with hydrogen electrolysers, themselves supplied with electricity produced by photovoltaic solar panels (510 kW) installed around the site (Bulletin Electronique USA, 287, 20/04/2012).

Although such devices seem to be well adapted for use in public transport, the cost of a personal vehicle equipped with a fuel cell remains too high to compete with systems based on the use of lithium accumulators, and, for the moment, vehicles using thermal motors[13].

This commercial handicap is mostly due to the fact that the electrodes have to be coated in catalysts made of precious metals, such as platinum. Active research is constantly looking into ways to minimize the quantity of precious metals used, or, better still, to replace them with less expensive electrode materials that are just as effective as platinum, which for the moment remains unrivaled as a holder of exceptional catalytic properties.

6.3.1. *Functional principles of PEMFCs*

A dihydrogen and dioxygen fuel cell operates on the basis of an oxidation reaction of the dihydrogen at the anode, which results in H^+ ions (protons), and a reduction reaction of dioxygen at the cathode in the presence of protons, described by the equation:

$$2\,H_2 - 4\,e^- \rightarrow 4\,H^+ \text{ (anode) and } O_2 + 4\,H^+ + 4\,e^- \rightarrow 2\,H_2O \text{ (cathode)}$$

13 Mercedes-Benz has been commercializing since 2010 a B-class (F-CELL) electric type car, a hybrid that follows on from the class A model that appeared on the market in 2004. It works with a fuel cell powered with dihydrogen (stored under a pressure of 700 bars) of 100 kW (about 136 HP) and a Li-ion battery of 35 kW and of capacitance 1.4 kWh that is charged using energy recuperated during braking. The current price is very high (about $100,000), but it should drop significantly in the next few years and come into the range of Diesel-Li ion battery hybrid cars. Comparable models are being developed by Peugeot-PSA, Honda, Ford and other automobile producers.

Overall, the transformation reaction is equivalent to the recombination reaction of hydrogen and oxygen ($2\ H_2 + O_2 \rightarrow 2\ H_2O$), the 4 H^+ formed at the anode being used in the reduction of O_2 at the cathode.

The membrane that separates the cathode and the anode plays a very particular role in the transformation process. On the one hand, it ensures the electrical insulation between the electrodes and stops any mixing of the gases, and on the other hand, it has chemical properties that allow only protons to cross. Its function is comparable to that of an ion-exchange resin.

In the present example, it is a Nafion™ membrane, initially produced by Dupont de Nemours, USA. They are fluorinated polymers, very chemically resistant, but very expensive, modified by sulfonate (SO_3^-) groups, able to exchange protons by guaranteeing their exclusive transport toward the cathode (Figure 6.16).

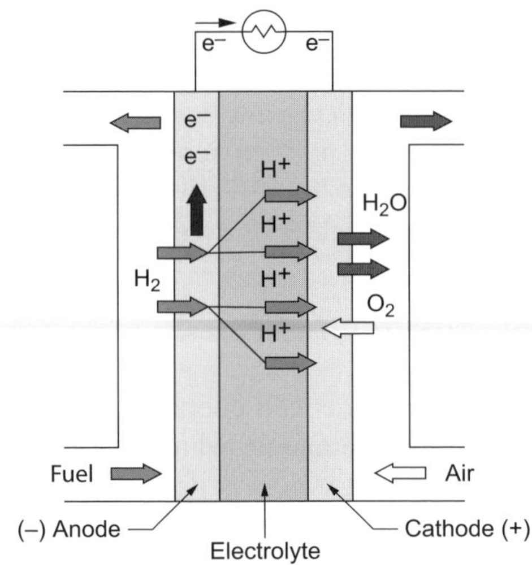

Figure 6.16. *Simplified diagram of a PEMFC using H_2/O_2 as a fuel. The protons produced by the oxidation of the fuel H_2 at the anode cross the membrane and react with the oxygen at the cathode according to the electrochemical reaction $4H^+ + O_2 + 4e^- \rightarrow 2\ H_2O$. The reaction produces heat and an electromotive force between 0.7 and 0.8 V (adapted from a Wikipedia document). For a color version of this figure see www.iste.co.uk/lacaze/nano.zip*

Each of the redox couples E°_{H+/H_2} and $E^{\circ}_{O_2,H^+/H_2O}$ imposes a redox potential on each electrode (Box 6.1) resulting in a theoretical maximal tension of the cell of 1.23 V in strong acidic conditions (pH = 0). In practice, as a result of loss caused by electrical resistance in the membrane, and also because the speed of the reduction of oxygen is not as fast as if the reaction was completely reversible, the tension of the cell does not go above 0.7–0.8 V; the density of the current is, however, very high, in the order of 1 A/cm^2 [WIN 04].

The electrical yield of the conversion is in the range of 50–60%, whereas theoretically, based purely on thermodynamic considerations, it should be 83%.

This is not very different from the concepts involved in classical electrochemical batteries or accumulators, except that the cell must be permanently supplied with fuel (dihydrogen), the oxygen being supplied by surrounding air.

The main advantage of fuel cells over accumulators is that they can perform continuously, with no decrease in performance, as long as the fuel supply is maintained. From a practical point of view, this is not a problem: in this respect, they can be considered equivalent to thermal motors, which keep working as long as the fuel tank is full. However, fuel cells do have a major flaw: their production is expensive, which greatly limits their potential exploitation commercially.

6.3.2. *Can the cost of dihydrogen fuel cells be reduced?*

The technology relying on the use of dihydrogen fuel cells with an electrolyte membrane in an acidic medium is currently the most reliable. Unfortunately, it remains very expensive. This is partially due to the inevitable use of precious metals as catalysts and partially due to the fact that the polymer electrolyte membranes that are the best adapted to use in acidic conditions are fluorinated membranes like Nafion™, a few microns thick and also very expensive.

The structure of these membranes is relatively complex. Each membrane comprises a main perfluorinated hydrophobic chain that has fluorinated branches with ether–oxide bridges that end in hydrophilic and ionizable sulfonated groups ($-SO_3^-H^+$), capable of releasing an H^+ ion (Figure 6.17). Hydrophobic and hydrophilic properties of the different fragments of the chain result in the nanostructuration of the membrane for a particular level of hydration, followed by the formation of aqueous nanochannels, a few nanometer wide. This considerably favors the diffusion of water and protons and highlights the excellent conductive properties of this type of membrane, especially at low temperatures (Schmidt and Chen [SCH 08]).

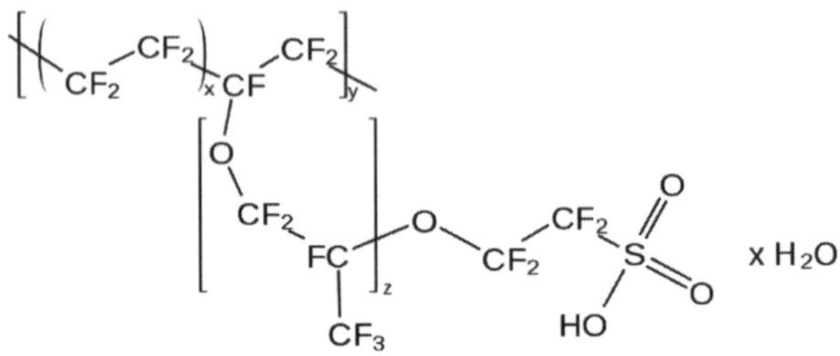

Figure 6.17. *Representation of a Nafion™ molecule carrying the ionophore group sulfonate SO_3H releasing protons by ionization*

The first idea that comes to mind to reduce the costs involved in producing polymer electrolyte fuel cells is obviously to reduce to a minimum the amount of precious metal (platinum), or even to replace it with a less expensive catalyst, but without significantly decreasing performances.

Research efforts have mainly been centered on increasing the speed of the reduction of oxygen in an acidic medium, where the presence of platinum on the cathode is a necessity. The question is: what is the lowest amount of platinum that can be deposited on the surface of the electrode without losing out on efficiency?

The oldest technique consists of dispersing platinum on carbon particles (carbon black)[14]. However, this technique is not ideal in that the dispersion of carbon particles on the membrane does not always guarantee good electrical interconnections between them, which decreases the yield of the oxygen's transformation reaction. This is the most critical issue and has been the subject of much important research over the last few decades.

The discovery of new carbon materials, graphene and CNTs, has sparked the launch of new paths of research. Indeed, important improvements in the speed of the reduction of oxygen have been observed when the commercial

---

14 The concentration of platinum spread out over the carbon black is in the order of 0.1 mg/cm^2 for the anode and 0.5 mg/cm^2 for the cathode [WIN 04].

Pt/carbon catalysts are replaced by "nanohybrids" composed of CNTs to which platinum NPs are attached.

But how can these improvements be explained? The answer is simple. The CNTs are good electrical conductors and therefore behave like electrical wire, which ensures that there is a very good contact between all the platinum NPs and the collecting electrode. This avoids any situation where the platinum NPs might find themselves electrically isolated from the current's collecting anode, which can happen when using traditional catalysts made from platinum NPs mixed with carbon particles (Figure 6.18).

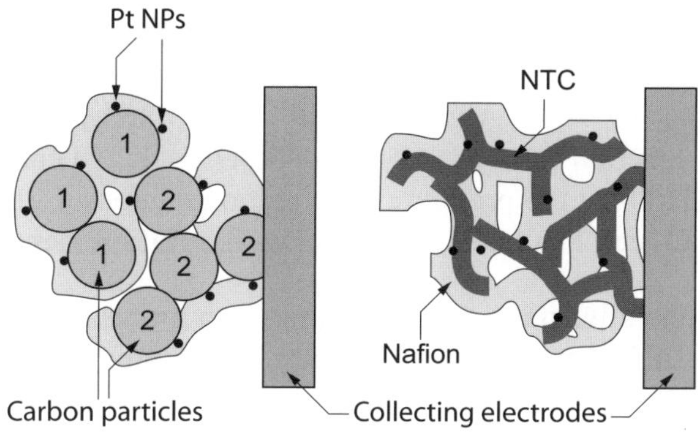

Figure 6.18. *Diagrams comparing the structures of catalyst layers made of platinum (Pt) NPs dispersed over carbon NPs or over CNTs, themselves coated in an electrolyte polymer like Nafion. The platinum is dispersed over the surface of the carbon NPs or CNTs (represented by black dots). Note that among the carbon NPs, those marked 1 are not connected to the electrode, unlike the NPs marked 2. The CNTs, comparable to conductive wire, are on the other hand all interconnected and directly or indirectly in contact with the electrode, which theoretically guarantees that all the electronic exchanges between O_2 and the platinum NPs adsorbed on the CNTs contribute to the current in the cell (adapted from Wei et al. [WEI 08])*

Another factor that contributes to the high cost of this type of cell is the Nafion™-type[15] fluorinated membrane, the success of which has yet to be surpassed. That said, it does have a weakness: it tends to become dehydrated when operating temperatures rise above 80 °C. This can be overcome by adding certain additives to the cell, which unfortunately slightly decreases conductivity.

15 To get a concrete idea of prices, a Nafion™ 211 membrane of 900 cm^2 without any catalysts can be bought for approximately $130–150. This high price is due to the fluoride technology that requires excessively costly equipment.

As a result of these considerations, other paths that do not imperatively require the use of fluorinated membranes have been envisaged. The alkaline domain is one of these alternatives.

The technology is certainly not new, and it is one of the cheapest technologies. The membrane has no particular function other than the diffusion of the gases throughout the solution, and stopping the cathode and the anode from forming a short circuit, while allowing the diffusion of the electrolyte, usually a solution of caustic potash (KOH).

The two electrode reactions are the same as for a cell in an acidic medium, except that they take place in an alkaline medium. Thus, at the cathode, the reduction of oxygen takes place following the reaction $O_2 + 2H_2O + 4e^- \rightarrow 4OH^-$, whereas at the anode, the hydrogen is oxidized following the reaction $2H_2 + 4OH^- \rightarrow 4H_2O + 4e^-$. This results in the formation of $2H_2O$, as the product of the recombination of $2H_2$ and O_2, just like for the H_2/O_2 cell in an acidic medium with a Nafion membrane.

Other than the advantage of functioning with an ordinary perfluorinated membrane (no sulfonate groups), the electrode reactions can take place using catalysts other than platinum (nickel and silver), which in turn considerably reduces the cost. The electrical characteristics are similar to those of an acidic electrolyte cell and can reach current densities of 1 A/cm^2 at 0.7 V with an electrical yield of approximately 50%.

These exceptional figures led to them being used in the Apollo space missions in the 1970s. However, despite this, there never was any form of commercial or industrial development.

A reason for this is that this type of cell works properly only if the oxygen is pure and exempt of any CO_2. An alkaline battery, constantly in contact with the surrounding air, is inevitably going to absorb some CO_2. Over longer periods of operation, potassium carbonate is formed, causing the deactivation of the catalysts as well as the obstruction of the membrane's pores, and consequently the dysfunction of the battery. This rules out any possible long-term use in mobile systems or electrical vehicles.

However, this inconvenience can be overcome. To do this, like for the PEMFCs, the use of a membrane that exchanges OH^- ions, for which the cation is no longer free to move around in the solution but is rather attached to a polymer chain, stops any possible formation of a salt such as K_2CO_3.

Such membranes do exist and are usually composed of linear chain polymers, to which are attached quaternary ammonium cations, $CH_2N^+(CH_3)_3$, which are associated to the OH^- anions and ensure in the solution the exclusive transport of the charges from the cathode toward the anode.

As a matter of fact, to operate satisfactorily in an alkaline H_2/O_2 battery, the membrane must present a combination of different properties comparable to those of the Nafion™, such as: (1) a high ionic conductivity (greater than 10^{-2} S/cm), (2) good mechanical resistance to be able to make membranes as thin as possible, (3) good chemical and thermal stability under temperature conditions of 100 °C or more and (4) if possible to be soluble in certain solvents, to be able to easily deposit them on electrodes by impregnation.

It was only recently, in 2008, that such membranes, of the polysulfone variety, connected by ammonium hydroxide groups (Figure 6.18), were first produced by a Chinese team [LU 08] (Professor Zhuang from Wuhan University). The latest improvement in synthesis engineering shows that these membranes have properties of ionic conduction, mechanical resistance and chemical and thermal stability that are similar to those of Nafion™, making them suitable to use in alkaline dihydrogen fuel cells [PAN 12].

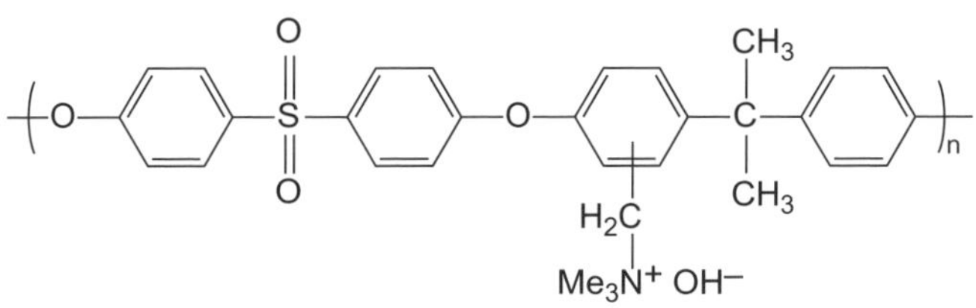

Figure 6.19. *Polymer in the OH^- ion exchanging membrane*

The first tests carried out with batteries with electrodes using catalysts other than platinum (silver cathode, nickel–chrome anode) look very promising. Although the power is still much lower than that of an H_2/O_2 fuel cell, already we can predict use in a multitude of different mobile devices, and possibly a launch into high-power alkaline batteries able to equip fully-electric vehicles in the next ten years.

6.4. General conclusions

The area of energy has recently become a societal problem closely linked to concerns about the environment. The production of "clean" energy has become a primary concern for our society.

On top of the development of alternative energy sources such as wind farms and hydroelectricity, the conversion of solar energy into electricity and the storage of this energy are two major problems that have particularly interested scientists.

Nanotechnology is particularly involved in both these domains. Concerning storage, the progress made with Li-ion batteries and supercapacitors is an example of this, with the increasing use of new nanocarbon materials that will probably change the storage capacities and will make all-electric vehicles more competitive, in terms of both cost and performance.

The use of hydrogen and oxygen in fuel cells is probably one of the best possible ways to create electric vehicles in the future. The only obstacle is the price, which is still too high to allow commercial development, especially in systems that use proton exchange electrolyte membranes that require the use of precious metals to catalyze the electrode reactions.

Another way is the use of OH^- ion exchanging membranes in an alkaline medium, as electrodes can be used without precious metals; the issue being finding an OH^- ion exchanging membrane that is sufficiently effective. Recent results obtained in the lab with basic electrolyte membranes with properties close to those of Nafion™ are encouraging. They should result in new synthesis research, which might provide the solution to the industrial development of alkaline batteries.

The conversion of solar energy into electrical energy is also a domain that is rapidly being developed. Competition between silicon and all-organic technologies has been stimulating, each possessing its own advantages and singularities. The goal of creating huge solar panels, made with techniques borrowed from the paints industry, is still far off but not impossible. This could mark a significant technological breakthrough in the production techniques currently in use and could ultimately radically change the commercial aspect of photovoltaic technology.

6.5. Bibliography

[BRO 00] BRODD R.J., HUANG W., AKRIDGE J.R., "Polymer battery R & D in the USA", *Macromolecular Symposia*, vol. 159, pp. 229–296, Wiley-VCH, 2000.

[BRO 01] BRODD R.J., Comments on the history of Lithium-Ion batteries, 2001, available at www.electrochem.org/dl/ma/201/pdfs/0259/pdf.

[BRO 04] BRODD R.J., BULLOCK K.R., LEISING R.A., MIDDAUGH R.L., MILLER J.R., TAKEUCHI E., "Batteries, 1977 to 2002", *Journal of the Electrochemical Society*, vol. 151, pp. K1–K11, 2004.

[BRO 12] BROUSSE T., TABERNA P.-L, FAVIER F., "Batteries et supercapacités", *Observatoire des micro et technologies*, pp. 196–201, 2012.

[BRU 12] BRUCE P.G., FREUNBERGER S.A., HARDWICK L.J., TARASCON J.-M., "Li-O2 and Li-S batteries with high energy storage", *Nature Materials*, vol. 11, pp. 19–29, 2012.

[DOE 07] Basic research needs for electrical energy storage, Office of Basic Energy Sciences, US Department of Energy (DOE), Washington, DC, 2007.

[DUN 11] DUNN B., KAMATH H., TARASCON J.-M., "Electrical energy storage for the grid: a battery of choices", *Science*, vol. 334, pp. 928–935, 2011, available at http://www.sciencemag.org/content/suppl/2011/11/16/334.6058.928.DC1/Dunn-SOM.pdf.

[EPR 10] Electrical energy storage technology options, Report 1020676, Electric Power Research Institute (EPRI), Palo Alto, CA, 2010.

[EPI 12] European Photovoltaic Industry Association, 2012, available at http://www.epia.org/index.php?id=18.

[GRÄ 09] GRÄTZEL M., "Recent advances in sensitized mesoscopic solar cells", *Accounts of Chemical Research*, vol. 42, pp. 1788–1798, 2009.

[GÜN 07] GÜNES S., NEUGEBAUER H., SARICIFTCI N.S., "Conjugated polymer-based organic solar cells", *Chemical Reviews*, vol. 107, pp. 1324–1338, 2007.

[HER 82] HERSCH P., ZWEIGEL K., *Basic Photovoltaic Principles and Methods*, Technical Information Office, US Department of Energy, 1982.

[IEA 11] Trends in photovoltaic applications. Survey report of selected IEA countries between 1992 and 2010, Report IEA- PVPS T1-20, 2011.

[LEW 06] LEWIS N.S., NOCERA D.G., "Powering the planet: chemical challenges in solar energy utilization", *Proceedings of the National Academy of Sciences of the United States of America*, vol. 103, pp. 15729–15735, 2006.

[LI 12] LI Y., "Molecular design of photovoltaic materials for polymer solar cells: toward suitable electronic energy levels and broad absorption", *Accounts of Chemical Research*, vol. 45, pp. 723–733, 2012.

[LIN 11] LINCOT D., "L'énergie photovoltaïque: les responsabilités nouvelles de la chimie", *Actualité Chimique*, vol. 353–354, pp. 52–60, 2011.

[LU 08] LU S., PAN J., HUANG A., ZHUANG L., LU J., "Alkaline polymer electrolyte fuel cells completely free from noble metal catalysts", *Proceedings of the National Academy of Sciences of the United States of America*, vol. 105, pp. 20611–20614, 2008.

[NAG 90] NAGAURA T., TOZAWA K., "Lithium ion rechargeable battery", *Progress in Batteries and Solar Cells*, vol. 9, pp. 209–217, 1990.

[NAO 12] NAOI K., NAOI W., AOYAGI S., MIYAMOTO J.-I., KAMINO T., "New generation 'nanohybrid supercapacitors'", *Accounts of Chemical Research*, 20 March 2012.

[PAN 09] PAN D., WANG S., ZHAO B., WU M., ZHANG H., WANG Y., JIAO Z., "Lithium storage properties of disordered graphene nanosheets", *Chemistry of Materials: A publication of the American Chemical Society*, vol. 21, pp. 31–36, 2009.

[PAN 12] PAN J., CHEN C., ZHUANG L., LU J., "Designing advanced alkaline polymer electrolytes for fuel cell applications", *Accounts of Chemical Research*, vol. 45, pp. 473–481, 2012.

[PHI 05] PHILIBERT C., The present and future use of solar thermal energy as a primary source of energy, International Energy Agency, Paris, France, 2005.

[PUM 11] PUMERA M., "Graphene-based nanomaterials for energy storage", *Energy and Environmental Science*, vol. 4, pp. 668–674, 2011.

[SAR 09] SARGENT E.H., "Infrared photovoltaics made by solution processing", *Nature Photonics*, vol. 3, pp. 325–331, 2009.

[SCH 08] SCHMIDT-ROHR K., CHEN Q., "Parallel cylindrical water nanochannels in Nafion fuel-cell membranes", *Nature Materials*, vol. 7, pp. 75–83, 2008.

[SIM 08] SIMON P., GOGOTSI Y., "Materials for electrochemical capacitors", *Nature Materials*, vol. 7, pp. 845–854, 2008.

[SUN 11] SUN Y., WU Q., SHI G., "Graphene based new energy materials", *Energy and Environmental Science*, vol. 4, pp. 1113–1132, 2011.

[WEI 08] WEI Z.D., YAN C., TAN Y., LI L., SUN C.X., SHAO Z.G., SHEN P.K., DONG H.W., "Spontaneous reduction of Pt(IV) onto the sidewalls of functionalized multiwalled carbon nanotubes as catalysts for oxygen reduction reaction in PEMFC", *Journal of Physical Chemistry C*, vol. 112, pp. 2671–2677, 2008.

[WIN 04] WINTER M., BRODD R.J., "What are batteries, fuel cells and supercapacitors?", *Chemical Reviews*, vol. 104, pp. 4245–4269, 2004.

[YAN 10] YAN J., FAN Z., WEI T., QIAN W., ZHANG M., WEI F., "Fast and reversible surface redox reaction of graphene-MnO_2 composites as supercapacitor electrodes", *Carbon*, vol. 48, pp. 3825–3833, 2010.

[ZHA 11] ZHANG Q., CAO G., "Nanostructured photoelectrodes for dye-sensitized solar cells", *Nano Today*, vol. 6, pp. 91–109, 2011.

Chapter 7

Nanobiology and Nanomedicine

7.1. Introduction

Over a century has passed since Paul Ehrlich[1], considered to be the father of chemotherapy, finished synthesizing the first ever drug designed to counter syphilis. The arsenic-based molecule (salvarsan) was remarkably efficient in treating the disease, which led to it being called a "magic bullet".

It was probably this notion of a therapeutic target that inspired researchers throughout the 20th Century, and led to the evolution of chemotherapy toward specific treatments, involving molecules with increasingly precise effects, the idea being to increase the selectivity of the drug toward the illness, and, at the same time, to protect it from a possible precocious breakdown using a variety of encapsulation techniques.

Since the 2000s, new strategies involving discoveries in the field of nanotechnology have been appearing in laboratories, and are on their way to becoming the medicine of the future, under the novel designation of nanomedicine.

---

1 Paul Ehrlich was a German doctor and physiologist (1854–1915), co-laureate in 1908 of the Nobel Prize (Medicine or Physiology) with the Russian scientist Ilya Ilyich Mechnikov, both for their work on immunology.

These new strategies rely importantly on the concept of targeted drug delivery, directing the drug toward a precise cellular target to treat a particular disease. This requires that the drug be associated with molecular structures that not only can play the role of markers so as to control in situ the drug's location within the organism, but are also able to recognize the specific molecules secreted by sick cells, and to bind selectively to them to deliver the drug.

This set of features is only a possibility if progress is made concerning the analytical methods available, including the miniaturization of the sensors, detection sensitivities able to pick up and dose chemical compounds at trace levels and the invention of new measuring devices able to follow the transfer of intracellular matter in situ; these new advances mark the start of a new area of research that allies nanoelectronics with biology, called bionanoelectronics. This area has recently become very important, and is now an intricate part of medical research.

Recent progress in nanotechnology is close to overcoming these different challenges. The objective of this chapter is to show this through several concrete examples that outline what can be hoped of from medicine and nanobiology in the 21st Century concerning the treatment of terrible diseases such as cancer, diabetes and cardiovascular afflictions.

7.2. Bionanoelectronics

The boundary that separates the world of the living from the outside has always been a source of fascination, and since the dawn of science researchers have tried to understand the mechanics behind the many functions that form the basis of living organisms. Biological systems are still greatly superior to anything that humans are able to create. To compare: the brain of a rat has a calculating capacity of 78 TFlops, while one of the most powerful computers in the world (IBM Blue Gene L), which has a volume of several cubic meters, has a calculating capacity of only 92 TFlops[2].

---

2 A TFlop (Teraflop) is equal to a thousand billion flops. The flop is an acronym that characterizes the execution speed of certain computer calculations. The capacity of the human brain is estimated at approximately 40 Petaflops (1 Petaflop is equal to a thousand TFlops). IBM announced in 2011 a new record of 100 Petaflops with the new generation of Blue Gene Q supercomputers.

One of the objectives of bionanoelectronics is indeed to create interfaces that would be able to accurately and instantaneously receive all the information involving the function of a living cell.

Protein–protein interactions, neuronal activity, the mechanisms involving the nucleic acids (DNA), and knowledge of ionic and molecular exchanges between the inside and the outside of a biological cell are the domains in which there remain many unknowns, and solving these requires powerful analytical methods.

The characterization and dosing of biological molecules at trace concentrations are also key points in detecting diseases. However, important information, both rapid and precise, needs to be acquired to be able to prescribe an adapted course of treatment.

We will try to show that such an objective can be achieved through the use of networks of multiplexed and transistorized NWs, which have the advantage of being easily accessible. Two significant examples are shown: the general dosing of proteins, in this case prostate-specific antigen (PSA), and the transmembrane ionic exchanges of a biological cell.

7.2.1. *The multiplexed detection of PSA using "transistorized" nanowires*

The use of semiconducting NWs integrated in a FET-type circuit is close to having a high enough level of sensitivity to allow the detection and dosage of a variety of molecules at very low concentrations.

A singularity of these FETs is that it is the molecules that are absorbed on the surface of the NW that function as the gate, and therefore are the causes of the variations in the NW's conductivity. Moreover, before receiving a signal, the surface of the SC must first be modified by particular receptor molecules that act as molecule "traps". It is this specific "receptor molecule-targeted molecule" interaction, typical of immunological assay, that is likely to induce variation in conductivity.

7.2.1.1. *Immunological assay of proteins by labeling*

One technique typically used consists of binding monoclonal antibodies (Ab) to the surface of the protein. These Ab are chosen for their specificity toward the target protein to be detected, and are associated with a specific antigen (Ag)[3].

The enzyme linked immunosorbent assay (ELISA) technique is an immunological protein assay invented in the 1970s that relies on the strong Ab–Ag bond, where the antigen Ag is the foreign protein that needs to be assayed.

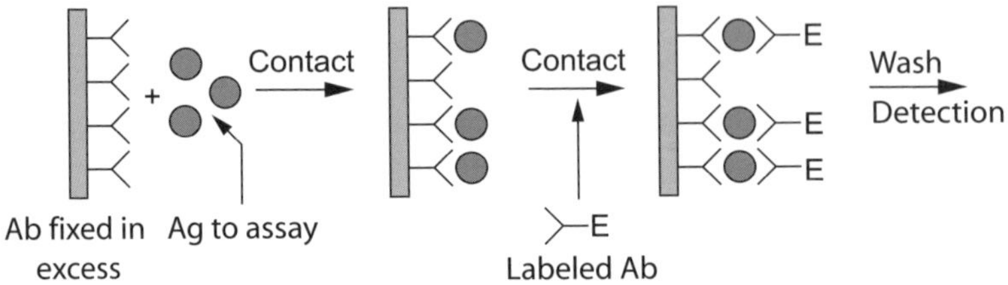

Figure 7.1. *Assay of an antigen (Ag) using an antibody (Ab) fixed on a solid substrate (ELISA technique, also called "sandwich" method). After the Ag binds to the Ab in excess, the Ag then binds to an Ab labeled with the enzyme E, which allows the colorimetric quantification of the Ag (adapted from a B. Media document, University of Paris 6)*

Among the many different variations of the technique, one of them consists of fixing the Ab on a solid surface; Ag is in solution and reacts with Ab on the surface. A second antibody, labeled by an enzyme, binds to the immobilized Ag. The enzyme reveals the location of the Ag, enabling the colorimetric or fluorimetric quantitative assay of Ag (Figure 7.1).

ELISA has been available commercially for many years, and entirely automated ELISA devices are present in all analysis laboratories.

3 Antibodies (Ab) are complex proteins secreted by an organism's immune system. They are formed of peptide chains roughly in a "Y" shape; antigens (Ag) are substances that are "foreign" to the organism and that induce the production of specific antibodies through an immune reaction. The antibodies can go on to bind to the antigen and neutralize it. The proteins secreted by an illness are seen as "foreign" by the organism and are therefore antigens.

7.2.1.2. *Use of nanowire networks*

Networks of multiplexed silicon NWs are easily constructed using the classical lithography techniques used in microelectronics (Figure 7.2).

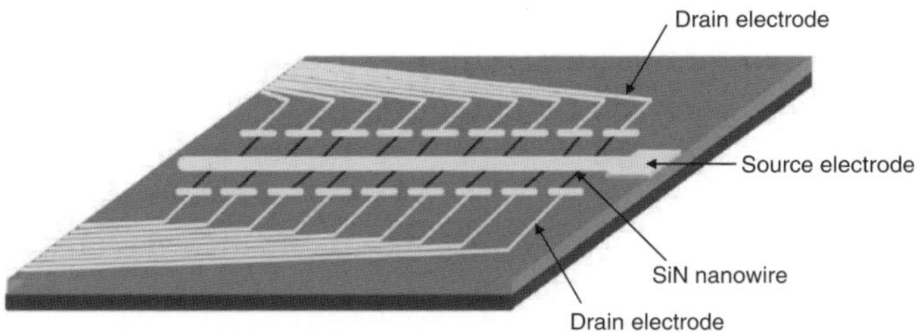

Figure 7.2. *Image of a network of nanowires made of silicon nitride (SiN) (adapted from Zheng et al. [ZHE 05]). For a color version of this figure see www.iste.co.uk/lacaze/nano.zip*

The surface of each NW is then modified by a different antibody, able to specifically bind a particular protein. This kind of assay is essentially immunological in nature, relying on the coupling of an antibody (Ab) to an Ag, but does not require the labeling of the reagents with fluoropores, radioactive elements or enzymes like in ELISA.

Figure 7.3 represents two NWs, each modified by a different antibody (Ab1 and Ab2), one of them being specific to a particular protein (Ag1).

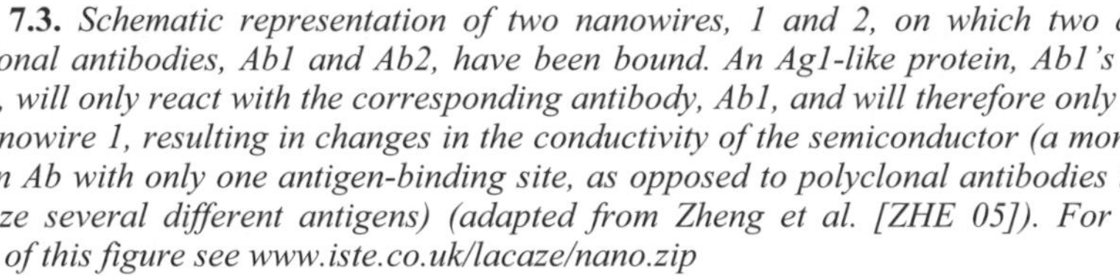

Figure 7.3. *Schematic representation of two nanowires, 1 and 2, on which two different monoclonal antibodies, Ab1 and Ab2, have been bound. An Ag1-like protein, Ab1's specific antigen, will only react with the corresponding antibody, Ab1, and will therefore only interact with nanowire 1, resulting in changes in the conductivity of the semiconductor (a monoclonal Ab is an Ab with only one antigen-binding site, as opposed to polyclonal antibodies that can recognize several different antigens) (adapted from Zheng et al. [ZHE 05]). For a color version of this figure see www.iste.co.uk/lacaze/nano.zip*

7.2.1.3. *The simplified and ultrasensitive detection of PSA*

PSA is a protein exclusively secreted by the prostate; the normal concentration of PSA in a man's blood is between 2 and 4 ng/ml. These figures increase dramatically in the case of prostate cancer; automated quantitative PSA assays, used in screening tests, are performed routinely using standard immunological techniques, usually with luminescence or coloring as the detection methods.

Compared to current assaying methods, used in all analysis laboratories and which function using specimen collection and manipulations with the help of machines, the device designed by Lieber *et al.* [CUI 01] has the advantage of being able to operate more simply, and consists of injecting the solutions to be assayed into a microfluidic circuit that directly irrigates the NWs through microchannels[4].

By injecting microquantities of liquid, alternately with or without PSA, into the microfluidic circuit, each change of the surface of the NW that corresponds to an Ab–Ag bond will appear as the change in the conductance of the NW.

This is exactly what happens when PSA is injected into the microfluidic circuit, and the NW's conductance is measured over time (Figure 7.4).

As expected, in the absence of PSA, the conductance of the NW stays the same and is maintained at a constant value (around 1,200 nS for the example in Figure 7.4), which corresponds to the conductance of the NW in contact with a neutral "buffer" solution.

As soon as the PSA is first injected (9 pg/ml or 9 trillionths of a gram per milliliter), and after a few minutes, the conductance increases dramatically, leveling off at a plateau of about 1,300 nS, which stretches over a period of roughly 800 s. Then, when the PSA-free solution again comes into contact with the NW, the conductance returns to its initial value of 1,200 nS, which means that the desorption of the PSA takes place when the surface is again irrigated by the buffer solution.

---

4 Microfluidics is a relatively new science that involves the development of systems designed to transport microquantities of liquid (between a billionth and a billionth of a billionth of a liter, or mathematically, 10^{-9} and 10^{-18} liter. A microfluidic circuit comprises a network of microchannels etched into the surface, the dimensions of which go from a few tens to a few hundreds of microns. While they remain commercially unexploited, the potential applications of microfluidics are huge and go hand in hand with the use of nanomaterials: they should begin to have an important part in biochemistry and biology, the pharmaceutical industry, and, according to Whitesides, should lead to a new generation of "labs on chips" [WHI 06].

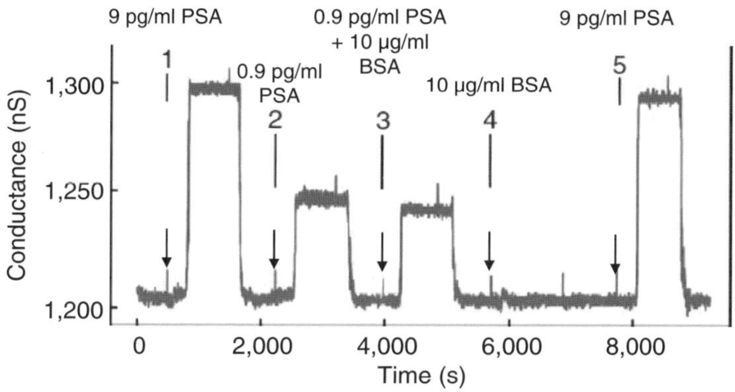

Figure 7.4. *Variation of the conductance of a nanowire in nanoSiemens (nS) as a function of the duration of the contact between the PSA and the SC. Each arrow (1, 2, 3 and 5) marks a different injection of PSA. Note the high reproducibility of signals 1 and 5, corresponding to PSA solution injections of 9 pg/ml, as well as the 0.9 pg/ml injections (2 and 3) that are not influenced by the presence of another protein, bovine serum albumin (BSA), without effect on the conductance of the nanowire (adapted from Zheng et al. [ZHE 05])*

Each time, there is a slight lag (several minutes) between the injection and the NW's response. This is characteristic of a microfluidic circuit, and corresponds to the time it takes for the reagent to reach the surface of the NW; in the same way, the return of the conductance to its initial value corresponds to the renewal of the buffer solution at the surface of the NW.

The same phenomenon happens again for repeated injections of PSA, with the conductance varying as a function of the concentration of PSA.

Furthermore, there is a notable specificity between the NW and its receptor: the injection of any protein other than PSA (BSA or bovine serum antigen) has no effect on the conductance of the NW.

The straight line that links the change in conductance to the concentration shows that there is a linear "semi-logarithmic" relationship with the concentration of PSA (Figure 7.5).

This logarithmic function implies that any increase in the concentration of PSA by a factor of 10 results in a constant increase in conductance of the same quantity (around 50 nS in the current example).

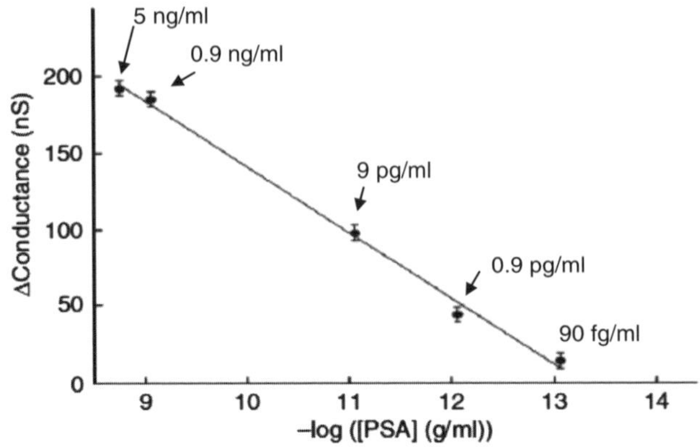

Figure 7.5. *Variation in the conductivity of a silicon nanowire (p-Si), modified by antibodies and in contact with solutions that contain different concentrations of PSA (adapted from Zheng et al. [ZHE 05])*

The ratio between the two extreme values of concentration, 900 pg/ml (0.9 ng/ml) and 0.09 pg/ml (90 fg/ml), is 10,000, which, on a logarithmic scale, is equal to a variation of four units, and therefore a change in conductance of roughly 200 nS. Clearly, this is a minute variation, but perfectly measurably with modern electronic devices.

The lowest value of 90 fg/ml is equivalent to a concentration of about 2 fM (femtomolar), meaning that the number of PSA molecules per milliliter is in the order of one million. This is not a lot!

The observation of these conductance variation cycles confirms that the NW measuring technique is well founded, and shows that three of the qualities that are most required in a detection device, sensitivity, specificity and reproducibility are absolutely present here. Such devices are well adapted to the production of analytical "chips". Being able to accurately and specifically determine the concentration of compound by a simple measure of conductance is a considerable progress over the traditional methods that require the labeling of the compounds by luminescent or radioactive groups, or even groups that color when in presence of a chemical revealing agent.

Once again, it must be pointed out that the very high sensitivity of the NWs to very small variations in the chemical composition of the surface is due to a property of all nanomaterials: the surface properties are much more significant than the volume properties.

Although this kind of technology is still at the experimental stage in laboratories, recent work shows that it can be easily applied to other areas of biology, notably in the simplified detection of DNA, which would lead to a new generation of DNA chips. Economically, this could lead to much more competitive prices, as the logistics of the analysis process would be much simpler and more readily available.

7.2.1.4. *Conclusions*

The use of semiconducting NWs in the detection and analysis of chemical or biological compounds at trace concentrations is a new technology that is taking off spectacularly. The required qualities of sensitivity, specificity and reproducibility are comparable, if not better, to those present in older, much more time-consuming techniques. Being able to selectively detect a compound and to establish its concentration using a simple measure of electrical conductivity is logistically highly advantageous. In the future, this should allow many complex devices to be replaced with portable systems that are much easier to operate, while still as efficient.

Currently, to the extent of our knowledge, the use of semiconducting NWs as receptors is still restricted to the laboratory. The large-scale production of a system that couples a microfluidic circuit to a network of NWs is perfectly feasible using the current methods employed by the microelectronics industry; the only problem is a chemical problem and involves finding distinct receptors that can be bound to the NWs and adapted to different targets.

7.2.2. *Connecting the organic and the artificial*

The boundary that separates the world of the living from the world of the artificial is extremely hard to overcome, especially at the level of a living cell. The introduction of ultramicroelectrodes in biology (only a few microns across) in the 1990s was a very important step forward. Developed in France and in the United States by Amatore and Wightman, they enabled the *in vivo* study of the diffusion of neurotransmitters between the synapses of nerve cells [AMA 11].

The advent of nanomaterials has considerably improved experimental methods; recently, there have been a large number of new studies on the scale of a single cell that have allowed direct communication between a living cell and the outside world.

This progress is due to new techniques that allow the production of nanosensors, smaller than cells, which are minimally invasive and can receive information that can describe the mechanisms of a biological cell.

Such experimentation has been performed by certain Californian teams (University of California at Los Angeles and Berkeley and the National Laboratory of Livermore). By employing FETs that use CNTs or semiconducting NWs as the sensitive detection components, the researchers showed that it is possible to visualize ionic exchanges in artificial cells, made up of lipid bilayers into which membrane transport proteins had been inserted.

7.2.2.1. *The construction of a nanosensor and its function*

The device is made up of a silicon NW (doped p) coated by a lipid bilayer, which is completely impermeable to any ionic exchange. Transport across the membrane can only take place through "access doors" that are made up of membrane transport proteins.

The lipid bilayer is very fluid; the insertion into the membrane of a protein like Gramicidin, essentially a spiral chain forming a channel, is easily done by simple impregnation. It is through this pore that the ionic exchanges take place, especially proton exchange (Figure 7.6).

The membrane is covered by a drop of water that contains the electrolyte to be tested (protons, calcium ions). The setup is different from a classic FET, in that the semiconductor covered by the membrane is in contact with the electrolyte, and that a reference electrode, immersed in the electrolyte, is responsible for the gate potential V_g.

The advantage of using a reference electrode is to be able to choose a gate potential that keeps the transistor in its most sensitive functional region (Figure 7.7).

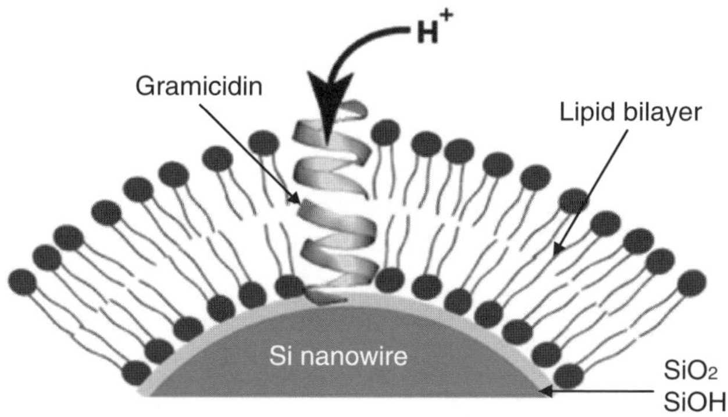

Figure 7.6. *Diagram showing the layout of the lipid membrane on the silicon nanowire. The nanowire is coated in a very thin insulating layer of silica, containing SiOH groups (not represented in the diagram), which are likely to transform into SiO^- groups when the solution becomes alkaline. The spiral-form Gramicidin is inserted between the lipid molecules and forms a channel in the membrane through which the proton exchange takes place (adapted from Misra et al. [MIS 09])*

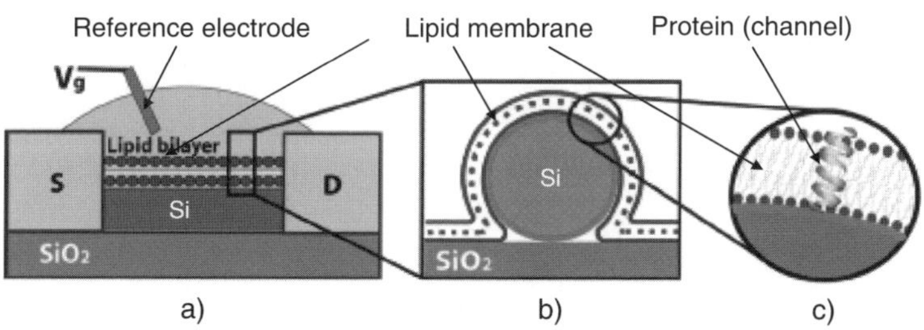

Figure 7.7. *Diagram showing the principle of the device used to measure ionic exchanges between the membrane and the surrounding electrolytic solution: a) longitudinal section of the "electrochemical" field-effect transistor showing the drop of electrolyte with the reference electrode (gate) maintaining a constant potential at the level of the nanowire; b) transversal section of the nanowire surrounded by the membrane in an omega shape; and c) formation of a channel by the insertion of a protein in the membrane (adapted from Misra et al. [MIS 09])*

The silicon NW's response to changes in the acidity of the solution (pH) is easy to understand. The surface of the silicon is slightly oxidized and has Si–OH groups (silanols analogous to alcohol functions) at its surface. When the medium becomes less acidic (pH[5] going from 5 to 7), the Si–OH groups

5 pH is a value that quantifies the acidity of a solution. It measures the concentration of H^+ ions, expressed on a logarithmic scale (a change of one unit of pH corresponds to a change of the concentration by a multiple of 10). Thus, pH 5 is equal to a concentration of 10^{-5} mol/l of H^+ ions, while pH 7 is 10^{-7} mol/l.

become the charged species SiO^-, which causes an electric field on the surface of the NW, increasing the conductance of the p-doped[6] silicon NW.

7.2.2.2. *Proton exchanges and their inhibition by calcium ions*

The curves of the changes in the conductance of the silicon NW represented in Figure 7.8 are particularly significant and once again confirm the high sensitivity of the device.

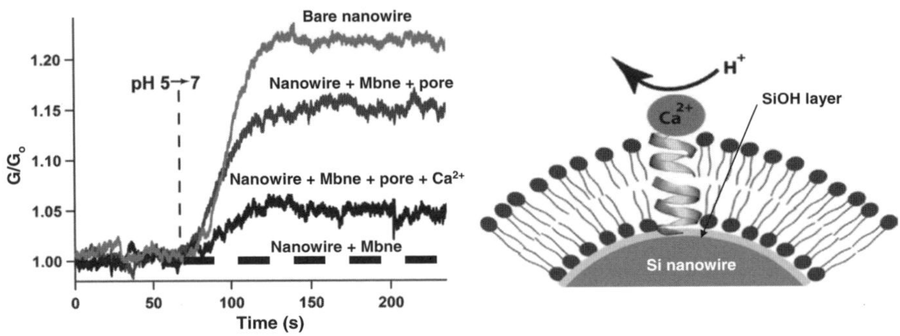

Figure 7.8. *Relative variation curves of the conductance (G/Go) of the silicon nanowire covered by a lipid bilayer, and irrigated by a solution with a pH between 5 and 7. The presence of Ca^{2+} ions in the solution results in the obstruction of the entrance of the channel and considerably decreases the flux of H^+ ions (adapted from Misra et al. [MIS 09])*

If the silicon NW is bare (not covered by a lipid bilayer), the variation in the pH from five to seven results in an important increase in conductance of more than 20%. On the other hand, when the NW is coated in a poreless membrane, the conductance remains unchanged when the pH of the solution is changed, which confirms its impermeability to protons.

As expected, the presence of pores in the membrane also leads to an increase in conductance, although this is less than the increase seen in the bare NW (around 15%) and therefore confirms that the exchange of protons can take place through the channel.

---

6 The source–drain (S–D) current is caused by the movement of the p-doped silicon's positive charges (holes). The electric field created by the negative charges (SiO^- groups) results in the increase in the concentration of the holes toward the surface, and therefore increase in the intensity of the current and consequently the conductance (the inverse of resistivity) of the NW.

However, the presence of large ions such as calcium can inhibit this exchange of protons by binding to the protein, which results in a new decrease of the conductance compared to the previous cases.

This type of behavior is entirely passive, which means that the protein plays no motor role in the transport of the H^+ ions. It is only the size of the pore that acts as a sieve and allows tiny ions such as protons to diffuse freely, while larger ions such as calcium are too big to enter the channel and in this way can block the entry.

7.2.2.3. *Conclusions*

The previous results confirm the soundness of the technique and can be easily applied to other types of membrane protein. An electrochemical transistor in which a reference (electrochemical) electrode is used instead of the traditional gate electrode of the FET is both unusual and original. Not only does this allow the selection of the transistor's optimal operating zone, but it also enables the setting up of a potential membrane control in a biological medium, and in this sense the observation of ionic pumps in action, pumps that are involved in the ionic exchanges of sodium and potassium, for example.

7.3. Nanomedicine

In the last decade, nanomaterials and nanotechnology have started to dominate medical research, leading to the branding of what is known as nanomedicine. Three major branches are the subject of intensive research: *in vivo* imaging and diagnosis, new therapies involving the controlled and targeted delivery of drugs and finally regenerative medicine, especially focused on the study of stem cells.

There is a lot of investment in this, and the pharmaceutical industry, because of the new treatment techniques, is due to increase this annually by 17% and is estimated at US$ 53 billion (Garcia *et al.* [GAR 07]).

Currently, cancer is one of the diseases that affect most people in the world. With a death rate of 12.5%, it is also one of the diseases for which most efforts are being made in the developed world to find a cure.

Although important breakthroughs have been made in the treatment of several cancers in the past few decades, some of the treatments currently used (chemotherapy and radiotherapy) are particularly traumatizing for the patient. As they are both invasive and non-specific, they also unfortunately cause the death of healthy and cancerous cells alike.

One of the major goals of nanomedicine is to change the therapeutic strategies and to invent new methods that would selectively treat the afflicted cells without harming the rest of the organism.

One of the first points to improve is in the detection of illnesses, which would involve the development of analysis techniques and *in vivo* imaging that would allow the detection of a disease at an early stage of its evolution, which is essential to increase chances of recovery.

The second point involves the treatment itself, with the idea of targeted drugs, able to reach a specific target without destroying the healthy cells in the immediate region surrounding a tumor.

7.3.1. *Biological barriers and the alteration of the cellular tissue surrounding a tumor*

A key point in making a diagnosis and determining the stage of development of a disease is being able to distinguish healthy tissue from sick tissue. For a decade now biologists and chemists have been developing new strategies based on the use of multifunctional NPs. The objective, after intravenously injecting the NPs, is that they selectively penetrate the tumor and bind to it, all the while sending out signals that can be detected and used to locate the afflicted region.

Fluorescence, magnetism or even the plasmonic and surface enhanced Raman spectroscopy (SERS) properties of NPs are currently used to do *in vivo* imaging. The biggest problem, however, is to think up ways that would avoid the diffusion of the marker toward regions other than the diseased region.

A developing tumor is deeply disruptive for the cellular tissue, which results in the dilation of blood vessels and the alteration of certain cells, which begin to secrete different proteins and enzymes in excess.

The lymphatic system is responsible for draining a variety of cellular waste and is also affected with reduced efficiency of the elimination process and strongly diminished permeability of the cellular tissue.

The dilation of the blood vessels and the decrease in the permeability of the cancerous tissue lead to a double effect known as enhanced permeability and retention (EPR). Not only does this cause the diffusion of the NPs toward the sick tissue, it also increases the retention of these NPs at the site of the tumor. These effects can be used to help selectively administer a drug or to image the afflicted region.

Before reaching their final destination, the NPs do, however, come across several biological barriers, which correspond to a variety of different biological defense mechanisms.

These could be a rapid opsonization[7] of the NP caused by proteins located in the blood plasma (opsonins) followed by the phagocytosis of the NP by macrophages (monocytes, neutrophils). The NPs can also be captured by the liver and the spleen, and can also be barred from penetrating the cancerous tissue, which is usually more reticulated than healthy tissue (Figure 7.9).

All these phenomena clearly oppose the functional efficiency of the NPs, a large part of which are thus condemned to rapidly disappear without ever reaching the biological target for which they were designed. In particular, they must escape the reticuloendothelial system (RES), which is made up of immune cells located in the liver, the spleen and bone marrow, and act after the opsonization of the NPs.

The final process can, however, be minimized by making the NP hydrophilic, especially by attaching polysaccharides (molecules that belong to the sugars family) or a particular polymer like polyethylene glycol (PEG) to the surface. This last change helps the NP hide from the immune system, and therefore considerably increases the circulation time.

---

7 Opsonization is essentially the coating of a foreign body by proteins. This makes the foreign body "visible" to the macrophages of the immune system (leucocytes, white blood cells) that are responsible for destroying them. Opsonization is a step that precedes phagocytosis.

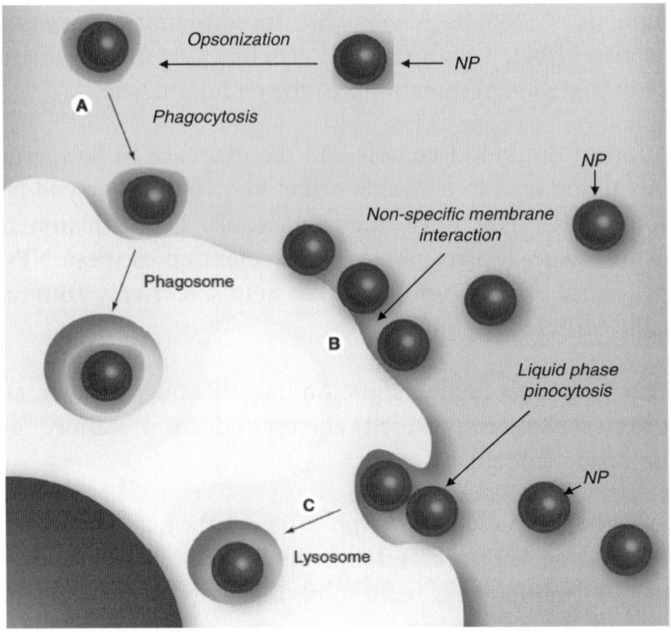

Figure 7.9. *Schematic representation of the different interaction mechanisms between the NPs and immune cells that are in blood plasma. Three mechanisms are described: the opsonization of an NP by proteins followed by phagocytosis (A); capture by immune cells caused by non-specific interactions (electrostatic or hydrophobic attractions) (B); and pinocytosis or a particular case of phagocytosis that corresponds to the absorption of an NP surrounded by liquid (C). Pinocytosis takes place especially with the cells that line the inside of the intestines (adapted from Nie [NIE 10])*

7.3.1.1. *The extravasation of nanoparticles toward cancerous tissue*

As indicated previously, the extravasation of the NPs is facilitated by the EPR effect, which results in the dilation of the vessels' pores at the level of the tumor, and in the accumulation of NPs in the tumor due to the poor draining of the NPs by the lymphatic vascular system[8]. This effect is optimally obtained with NPs a hundred nanometer in diameter, which are

---

8 A tumor can be considered the result of an anarchic growth of cells that escapes the control of the organism. This uncontrolled growth is caused by certain proteins, called epidermal growth factor (EGF). The protein is located outside the cell and can bind to the corresponding receptor EGFR (epidermal receptor of the growth factor located in the cellular membrane) and thus set off a signal that is transmitted to the cell nucleus and induces permanent cell division. Other receptors are also prone to occasional overexpression by a tumor: the transferrin receptor (iron regulator), the folate receptor (folic acid regulator), overexpressed in ovarian cancer.

dimensions that are compatible with the dimensions of the pores, allowing them to selectively travel toward the affected tissue (Figure 7.10).

The diffusion of the NPs can either be passive (simple NPs or QDots) or selectively directed toward cancerous cells. In this last case, the NPs are modified by molecules that will spontaneously bind to receptors that appear massively on cancerous cells. These receptors are quite few in healthy cells, while in cancerous cells they multiply (they are "overexpressed": there is an overproduction).

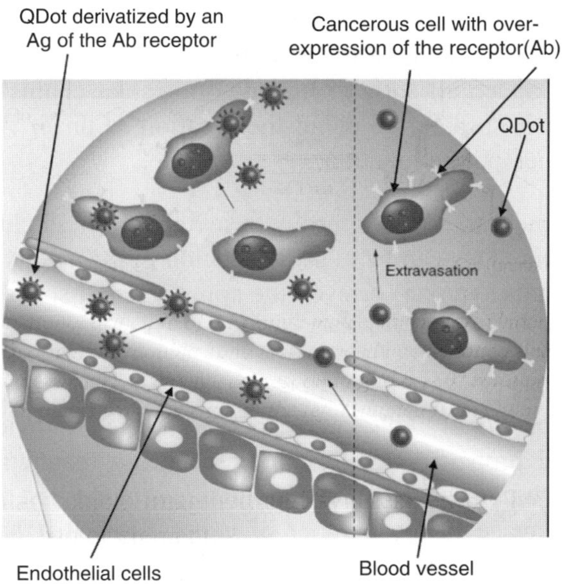

Figure 7.10. *Mechanism of the surface diffusion of functional and non-functional NPs by an Ag of the Ab receptor (represented by 'Y' on the cancerous cell), bound to the cellular membrane. The region located above the vessel corresponds to the tumor cellular region, the region below being the healthy region. The presence of abnormally large pores on the side of the tumor allows the extravasation of a large amount of the NPs toward the cancerous tissue. These NPs can go on to specifically bind to the receptors (Y) of the cancerous cells after having been modified by molecules that have a high affinity for these receptors (adapted from Nie [NIE 10]). For a color version of this figure see www.iste.co.uk/ lacaze/nano.zip*

It is the antigen–antibody combination between membrane receptors and NPs that is used in imaging and especially to deliver medication selectively at the level of the tumor.

7.3.2. *Nanoprobes for in vivo real-time imaging*

The use of nanoprobes in the shape of multifunctional particles is a promising concept for imaging and medical therapies.

Several favorable elements justify this: (1) their size, which can be modulated between 10 and 500 nm, is far less than that of a cell, which is of several microns; (2) the pores of blood vessels in a region affected by a tumor are just the right size that allows NPs of roughly 100 nm in diameter to diffuse across, and therefore to irrigate the sick region; (3) capture by the kidneys is less than for simple molecules or small NPs, and the RES in the liver absorbs them less than particles of sizes in the order of the micron; and (4) it is easy to attach all sorts of ligands (molecular chains with particular functions) to their surface, allowing the creation of an *in vivo* image, and also to bring a therapeutic function.

The development of a tumor results in several different chemical changes in its environment. This can be a local variation in the acidity, lower oxygen levels than normal (hypoxia) or even the appearance of specific proteins and enzymes.

The high levels of proficiency that have been reached in the synthesis of various multifunctional NPs and the complete knowledge of their physicochemical properties (SPR, luminescence or magnetism) have seen the rise of new strategies of imaging and therapy. In the past few years, there has been an increase in the number of experiments that have proven the efficiency of these techniques, which are repeatedly the subject of reviews and improvements; the following section shows several examples related to certain pathologies.

7.3.2.1. *Imagery resulting from plasmon resonance of gold NPs and from their interaction with enzymes characteristic of a pathological process*

Previously, we have seen that the color of gold NPs varies considerably depending on their state of aggregation (see Chapter 4, Figure 4.3).

Such a property was recently utilized by several Californian teams to reveal *in vivo* a phenomenon of apoptosis, important in cellular reproduction and vital to oppose the anarchic growth of cancer cells.

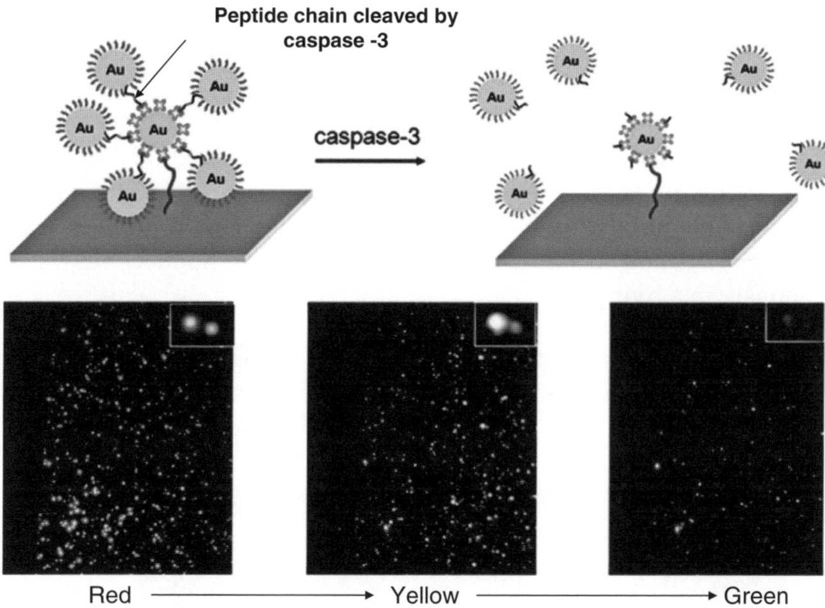

Figure 7.11. *In vitro observation under the light microscope of the color of gold NPs bound together by a peptide chain, degradable by caspase-3. The color goes progressively from red (group of five to six NPs, usually about 40 nm) to green (isolated NPs, the smallest being 5–10 nm in size) due to the sectioning of the peptide chains caused by caspase-3 (adapted from Jun et al. [JUN 09]). For a color version of this figure see www.iste.co.uk/lacaze/nano.zip*

A particularity of apoptosis is that it is accompanied by the production of certain enzymes (proteases), which have the property of cutting certain peptide chains (assembly of amino acids)[9].

By using gold NPs as reagents (five to six gold NPs linked together by peptide bonds), a change in coloration and intensity can be induced in the cells caused by the action of proteases that are characteristics of apoptosis.

9 Apoptosis is a biological process that is genetically controlled and ends in the natural death of cells of the organism, without inflammation, as opposed to the death of cells by necrosis or senescence. It plays an important role in morphogenesis as a factor regulating cell growth. Several diseases, such as AIDS, cancer and other neurodegenerative pathologies, involving the anarchic growth of cells, are thought to be caused by a defect in the apoptosis process.

By cutting the peptide bonds that link the gold NPs an enzyme such as caspase-3[10] causes a change in coloration from an intense red (aggregated NPs) to low intensity green (isolated NPs). (Figure 7.11).

The main advantage of using gold NPs is their great stability and the fact that they are not toxic, which enables their use in long experiments. Because of this technique, the reinitialization of the apoptotic process can be followed *in vivo* and in real time when an apoptosis inducer[11] is injected intravenously in a living cell.

Such experiments have been carried out on living cells in real time by Californian teams. Figure 7.12 shows at a single instant a map of the intensity of the light diffused by the gold NPs that have penetrated the cell.

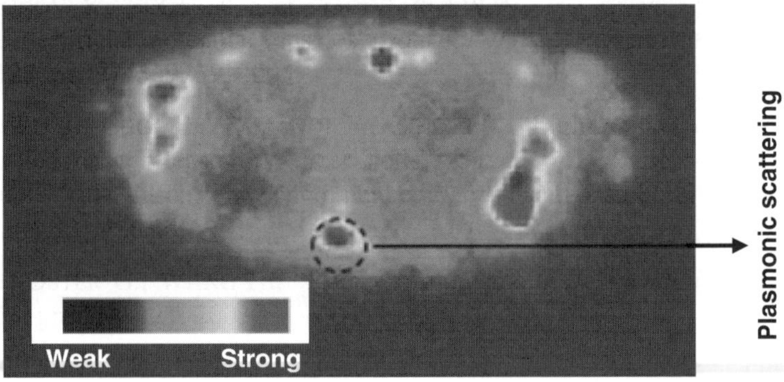

Figure 7.12. *Light microscope image of a living cell in which an apoptosis inducer and gold NP aggregates (same as in Figure 7.11) have been injected intravenously. The dark gray stain in dotted circle (dark red on Web) represents the plasmon emission of an aggregate of gold NPs that is still not dissociated. The very pale gray around the dotted circle (yellow area on Web), corresponding to weak intensity, is related to aggregates that have lost one or more NPs, while the medium gray (pale blue area on Web), corresponding to a low intensity, represents the light emitted by isolated gold NPs, resulting in the destruction of the aggregates by the protease. The deep gray area (strong blue on Web) around the cell correspond to a very low intensity of the scattered light (adapted from Jun et al. [JUN 09]). For a color version of this figure see www.iste.co.uk/lacaze/nano.zip*

10 Caspases are enzymes that are able to cleave proteins and therefore peptide bonds. They are responsible for the process of apoptosis and therefore restrict the anarchic proliferation of cells.

11 An inducer of apoptosis sets off the mechanism of apoptosis in a living cell, which in turn will secrete a protease such as caspase-3. In a cancerous cell, which is reproducing in an anarchic fashion, the apoptotic function is absent, and the injection of an apoptotic inducer is currently a path explored as a way of fighting the tumor.

The technique also allows us to follow, as a function of time, the light scattered by NP aggregates and their gradual disappearance under the effects of caspase-3 (Figure 7.13). The induction time marked on Figure 7.13 corresponds to the time that has gone by between the intravenous injection of the apoptosis inducer, the penetration of the inducer into the cell by endocytosis[12], and the production of caspase-3 by the cell, leading to the progressive breakdown of the NP aggregates.

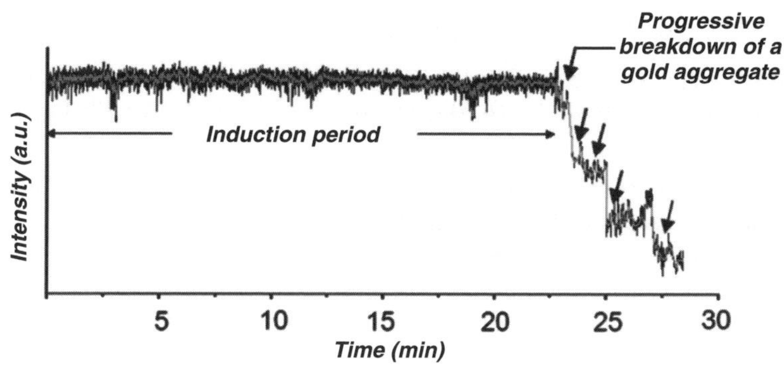

Figure 7.13. *Evolution of the intensity of the plasmon emission of an aggregate of gold NPs between the moment of the injection of the apoptotic inducer (AI) and the progressive breakdown of the aggregate following the action of caspase-3, induced by the AI after penetration in the cell (the arrows on the graph correspond to the successive losses of NPs, in the initial crown) (adapted from Jun et al. [JUN 09])*

7.3.2.2. *Luminescence imaging triggered by enzymes or reactive oxygen species characteristic of a pathology*

The technique, in this case, relies on the emission of luminescence, also caused by proteins or enzymes, secreted by a living cell in response to a dysfunction within the cell.

The principle is more or less the same as before, using gold NPs carrying peptide chains that can be broken down by proteases or other specific molecules that are characteristic of a particular pathology.

The difference with the previous case is that the chains contain fluorophores (chemical fluorescent groups) that emit infrared light when

12 Endocytosis is a process during which the cellular membrane absorbs a particle by forming an invagination and enveloping it in a membrane vesicle. It allows the cell to internalize into its own cytoplasm an element that was previously in the external biological fluid.

they are illuminated[13]. This fluorescence (see Box 3.1) is actually "turned off" by the gold NP and is only "turned on" when the fluorophores are separated from the NP and are released into the surrounding medium.

The property, which is characteristic of gold NPs, corresponds to the fact the energy acquired by the fluorophore during illumination is directly transferred to the NP (energy transfer or ET), which annihilated the fluorescent emissions (extinguishing phenomenon known as "quenching").

In the example of Figure 7.14, the chain carrying the fluorophores is hyaluronic acid,[14] bound to the gold NP through a sulfur atom (S) according to the classical technique of functionalization of gold NPs (see Chapter 3).

This technique has proven to be particularly efficient in locating certain pathologies within an organism, such as arthritis, which is usually accompanied by "oxidative stress".

This biological concept, used to describe a state of imbalance in the organism, corresponds to an excess of reactive oxygen species (ROS), which is highly dangerous because they are likely to react with a large number of biological molecules such as DNA, proteins and lipid bilayers.

These ROS are small molecules like hydrogen peroxide (H_2O_2), but most often a variety of radical species (molecules or ions comprising a non-paired electron), such as hydroxyl radicals ($^{\bullet}OH$), peroxide ion ($^{\bullet}O_2^-$), nitrogen monoxide ($^{\bullet}NO$). They are responsible for ageing and many diseases including cancer, cardiovascular diseases, atherosclerosis and arthritis and thus, can be used as biological markers for these diseases.

Due to their very high reactivity, they are able to fragment hyaluronic acid, and therefore, the scheme of Figure 7.14, which describes the emission of fluorescence, can be applied in the *in vivo* detection of ROS generated by a cancerous tumor or arthritis.

---

13 Fluorophores are chosen to emit near the infrared region of the spectrum (wavelengths between 650 and 900 nm), which allows penetration of the cellular tissue several centimeters deep, which is an important advantage in *in vivo* imaging.

14 Hyaluronic acid is a natural compound found in the cornea, skin and cartilage. It is a thickening agent, used in plastic surgery to get rid of wrinkles, and also as a lubricant for joints in the case of rheumatoid arthritis.

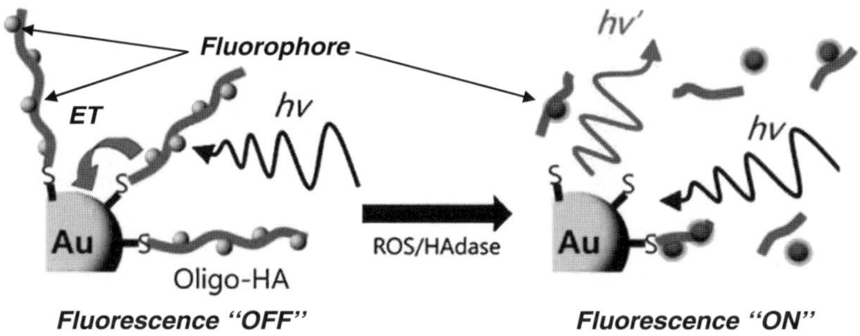

Figure 7.14. *Gold NPs bound by oligo-HA (hyaluronic acid) chains carrying fluorophore groups (gray sphere(s)). Under the effect of infrared light (hv), the fluorophores transfer the energy received to the gold NP (ET); there is no emission of fluorescence (Fluorescence "OFF"). In the presence of cleavage agents, which can be reactive oxygen species (ROS) or particular enzymes (HAdase), the fluorophores are separated from the NPs and released into the surrounding medium. The transfer of energy toward the NP being no longer possible, the fluorophores fluoresce (hv') (adapted from Koo et al. [KOO 11])*

Such experiments were carried out by a Korean team on mice, which had previously been made arthritic three weeks before the imaging experiment (inoculation intradermic of specific drugs that generate arthritis in the joints of a mouse's paw).

In the case where the gold NPs are administered intravenously, an important fraction of the NPs is picked up by the endothelial cells of the liver. Also, to minimize the extent of this undesirable capture, an injection of molecules able to saturate the hyaluronic receptors is done 30 min prior to the injection of the gold NPs.

An emission of fluorescence in the left paw of the mouse can then quite clearly be detected three hours after the injection of the gold NP aggregates. Its intensity reaches its maximum after six hours, and decreases abruptly 24 hours after the injection (Figure 7.15). The concentration of "hyaluronic" NPs is very low (in the order of a few picomoles, or one ten thousandths of a milligram), which confirms the high sensitivity of the technique, which effectively enables the detection of ROS at trace concentrations.

7.3.2.3. *Magnetic resonance imaging coupled with nanophototherapy*

Magnetic resonance has become a routine radiological medical imaging technique, used increasingly in diagnosis (magnetic resonance imaging or MRI).

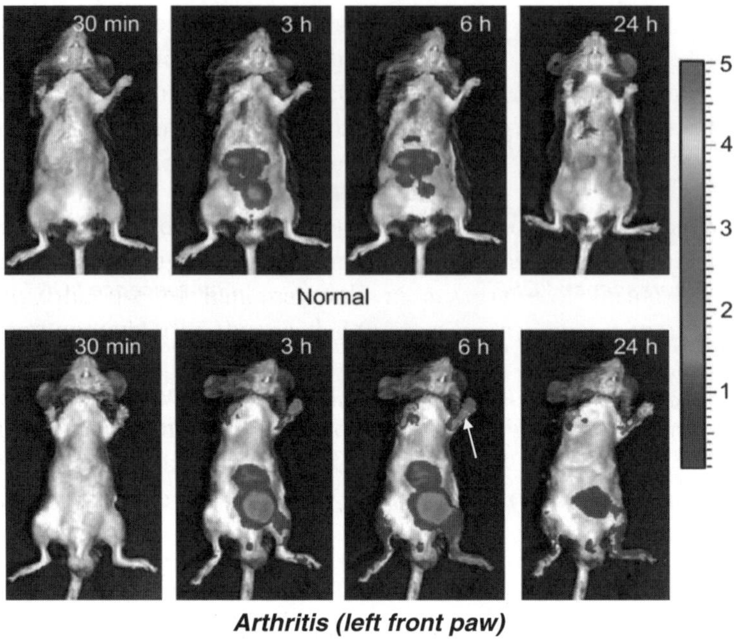

Figure 7.15. *Comparative imaging between a normal mouse's paw and a paw suffering from arthritis after intravenous injection in the mouse's tail of a solution of gold NPs bound to chains of hyaluronic acid modified by fluorophores. The highest intensity of infrared fluorescence (red area in the Web version), which indicates a highest concentration of ROS appears after six hours (white arrow, bottom image) and decreases only slightly after 24 hours observation (yellow area in the Web version). It can be noted that for the normal mouse (top images) the ROS are undetectable in the paws of the mouse. The intensity of the fluorescence is very weak in the areas of the liver (blue areas in the Web version), due to a previous saturation of the hyaluronic receptors of the liver by specific molecules (adapted from Lee et al. [LEE 08]). For a color version of this figure see www.iste.co.uk/lacaze/nano.zip*

The signal measured is directly linked to the density of the number of hydrogen atom nuclei, mainly belonging to water molecules, present in the cellular tissues. This allows a clear distinction between the different organs of the human body, with important image contrasts, sufficient to differentiate between a tissue affected by a tumor and a healthy tissue.

The technique uses very low electromagnetic radiation energies (radio frequencies), which have no effect on the organism, making it perfectly safe. Its only defect is that it is a very heavy and extremely costly (both to buy and to run) installation, which limits its presence to large medical centers.

When a very precise analysis of organs is necessary, contrast agents are injected intravenously. For this, paramagnetic compounds of gadolinium ions (Gd^{3+}) are used, which result in an increase of the intensity of the signals of hydrogen atoms relative to the water molecules. Unfortunately, they are very toxic compounds, and they need to be coated in particular complex molecules (crown molecules), that isolate them from the biological medium, making them a lot less toxic, but without losing their contrast agent properties.

This toxicity of Gd^{3+} contrast agents is not, however, completely removed; a new trend involves replacing the gadolinium with hybrid ferromagnetic NPs (core–shell iron–gold magnetite or Fe_3O_4 @Au in which Au is the shell and Fe_3O_4 the core), which are non-toxic, and are also considered to be very good contrast agents.

This new technique, which appeared recently is developing rapidly, and benefits from progress made in the elaboration of multifunction hybrid NPs (HNPs). This type of NP can locally destroy cancerous cells using photothermic methods, all the while creating an imaging through nuclear magnetic resonance. Laser irradiation of the hybrid NP in the near infrared has the advantage of deeply penetrating the tissue without damage, and by plasmon resonance to cause the heating of the gold shell (40–50 °C) of the NP, enough to induce apoptosis in the cancer cell.

By using techniques of NP functionalization as a way of targeting cancer cells, we can hope to selectively destroy them without changing the healthy cells.

These very effects were recently demonstrated by an American university team (Cornell University, Ithaca) in the case of two stains of colorectal cancer cells (SW 1222 and HT 29), treated with Au–F_3O_4 hybrid NPs.

The experiment is performed *in vitro*. To make the HNP target the cancer cells, an antibody (scFv) that antagonizes the antigen (A33) is bound to the NP. The antigen A33 is only secreted by the strain SW 1222, which implies that the targeting effect of the HNPs can only be obtained with the SW 1222 strain. After laser irradiation in the near infrared (λ = 808 nm), the cells are marked by a particular colorant (calcein AM), which emits a green fluorescence if the cell is alive.

The microscope images of Figure 7.16 show that the cells targeted by the HPNs have been destroyed (black circle) by a low irradiation intensity (5.1 W/cm^2, Figure 7.16(b)), while the non-targeted cells, which contain very little HNP, are only destroyed by much higher intensities, in the order of 31.5 W/cm^2 (Figure 7.16(h)). This shows the heat concentrator effect of HNPs that, for low irradiation energy, cause the destruction of cancerous cells without touching the healthy ones.

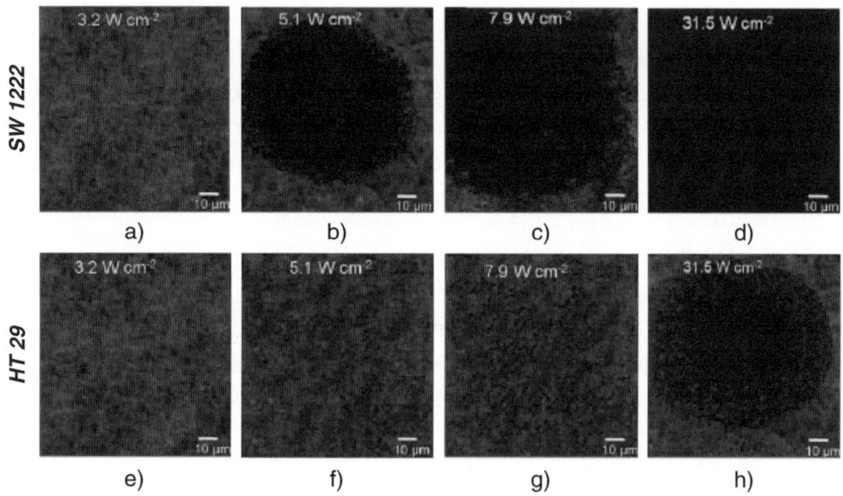

Figure 7.16. *Comparison of the effect of incubation of two lines of colorectal cancer cell SW 1222 and HT 29 by HNPs able to target only the SW 1222 strain. Both strains were irradiated in the near infrared over 6 min with increasing irradiation intensity (3.2, 5.1, 7.9 and 31.5 W/cm^2). The marking by calcein AM after irradiation shows the death of the SW 1222 cells (black circle) starting at 5.1 W/cm^2, while the cells of the HT 29 strain that did not bind the HNPs resisted necrosis until 31.5 W/cm^2 (adapted from Kirut et al. [KIR 10]). For a color version of this figure see www.iste.co.uk/lacaze/nano.zip*

7.3.2.4. *An innovative strategy for an improved penetration of the NPs in the cancerous cell tissue*

It has been previously shown that the optimal size of an NP used to create an *in vivo* image of a tumor is in the order of a hundred nanometers, which allows it to diffuse across the pores of hyperdilated blood vessels and to accumulate around a tumor, principally as a result of the EPR effect.

The problem is that this size stops the NPs from penetrating deeper into the cancerous tissue, which is made of very dense collagen. In these conditions, therapeutic treatment is unlikely to be successful if the administration of medication is done with this type of NP and if the tumor has already reached a stage of metastasis.

A primary issue is to find a way that would allow the NPs to penetrate more deeply into metastatic cancerous tissue, both in imaging and therapy.

A solution was recently suggested by several researchers on the East coast of the United States (Massachusetts Institute of Technology, Massachusetts General Hospital and Harvard Medical School). The innovative concept involves synthesizing NPs with dimensions that self-adjust depending on the physiological barriers encountered after their extravasation out of the blood vessels.

This requires the synthesis of NPs with an initial size of roughly 100 nm that would allow normal extravasation outside of the blood vessels, and then that this size be reduced to roughly 10 nm to allow their diffusion into the tumor tissue.

Such an effect can only be obtained if the chemical species secreted by the tumor are able to cause the breakdown of the NP and lead to its fragmentation into smaller NPs.

The solution put forward by Popovic *et al.* [POP 10] is particularly ingenious and relies on the synthesis of a 100-nm gelatin NP, containing a small number of QDs roughly 10 nm in diameter each. This gelatin is dissolved when it comes in contact with proteases (enzymes abundantly secreted by the tumor), freeing the QDs. These can then easily diffuse in the tumor tissue and apply their function as a biological marker or as a transport vector for medication in the very heart of the tumor (Figure 7.17).

The concept was recently validated and comparative tests between NPs of comparable size, both degradable and non-degradable, were performed *in vivo* in mice presenting a tumor. Penetration differences are clear between the two NP categories (Figure 7.18).

The above presentation of the process is obviously very simplified, and omits many experimental details. They are in fact a highly sophisticated experiment that can only be carried out by the perfectly coordinated participation of chemists, biologists and physicists.

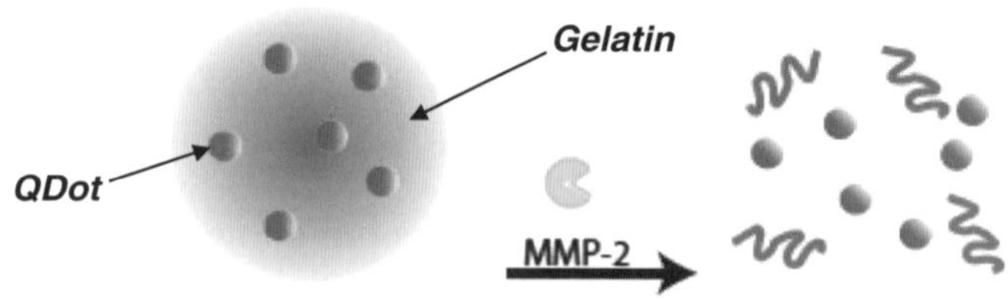

Figure 7.17. *Principle behind the size change of a gelatin-QDs composite NP. The initial size of the NP is roughly 100 nm; under the action of a protease (gelatase MMP-2), the gelatin is dissolved and frees the QDs that have a diameter of 10 nm (adapted from Wong et al. [WON 11])*

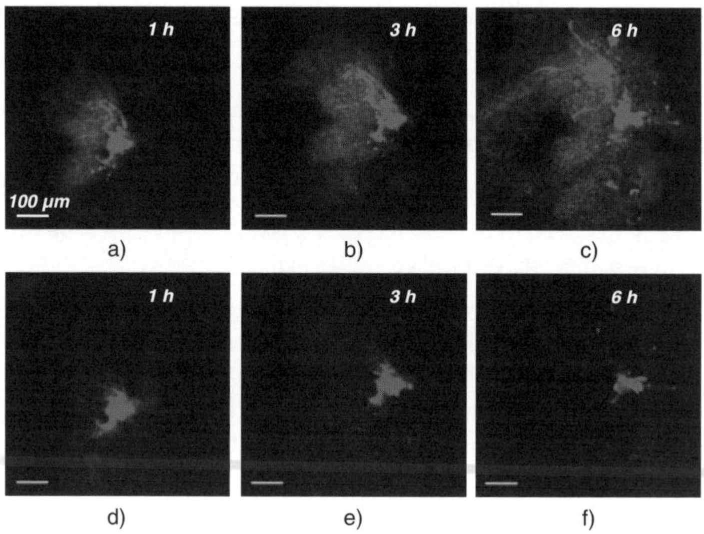

Figure 7.18. *In vivo images of fluorescence from degradable (a, b, c) and non-degradable (d, e, f) NPs, coinjected into cancerous tissue and obtained 1, 3 and 6 h after injection. The NPs are all the same initial size (roughly 100 nm). The degradable particles free the QDs (10 nm) that emit a green light (a, b, c) while the non-degradable NPs (QDs of about 60 nm) protected by an inert crown of silicon emit a red light (d, e, f). Six hours after injection, the gelatinized have been largely distributed into the cancerous tissue (circle over 300 µm in radius compared to the center of the injection) while the non-degradable NPs stay located at their injection center without any notable diffusion (adapted from Wong et al. [WON 11]). For a color version of this figure see www.iste.co.uk/lacaze/nano.zip*

This requires that several biological verifications be made, necessitating advanced skills in the synthesis and functionalization of NPs, and also profound knowledge of the mechanisms involved in the circulation and retention of NPs in the vessels.

It is of particular importance that the NPs created be non-toxic, and that they have a circulation time in the blood vessels that is long enough to allow imaging over several hours. At the same time, it is also absolutely vital that they could be removed from the organism in a reasonable amount of time.

These are recent experiments and are still far from clinical level experimentation. The next step would be to check that effective therapeutic improvements are brought by the use of degradable NPs, capable of delivering medication likely to diffuse into the whole of the tumor in a perfectly targeted manner.

7.3.3. *Challenges of nanomedicine and some significant clinical results*

The use of nanodrugs injected intravenously has become a common practice in nanomedicine. Important results have been obtained and several drugs used to treat tumors have had their efficiency improved, as a result of changes at a nanometric level and by using the targeting effects of certain ligands.

7.3.3.1. *The first commercial nanomedication*

In the 1990s, American and Israeli teams showed that liposomes loaded with doxorubicin[15] and modified on the surface by hydrophilic polymer chains (PEG) were remarkably efficient in the treatment of certain tumors [PAP 91]. This improvement, caused principally by an increase in the time of retention in the vascular system, a result of the presence of PEG derivatives at the surface of the liposome, led a bit later to the commercialization of Doxil®, a nanodrug corresponding to a new formulation of doxorubicin, authorized in the United States from 1995 onward.

The targeted release of a drug into the heart of a diseased tissue marks a new step in the design of nanomedication associated with transport "nanovectors". The strategy is comparable to the one described for *in vivo* imaging and comes down to injecting multifunctional NPs containing ligands able to selectively bind to membrane receptors that are overexpressed in the disease, into the venous cardiovascular system.

15 Doxorubicin is a drug used in the treatment of several cancers; it acts on DNA of cells by blocking its replication, which leads to the death of the cell.

Taxol®, prescribed in the treatment of breast cancer, is one of the first examples to illustrate this orientation toward "intelligent" nanodrugs, designed to be active medication.

In its initial formulation, Taxol® corresponds to a mixture of paclitaxel[16] (very hydrophobic) and of surfactants (Cremophor – EL®, a synthetic derivative made by BASF and extracted from castor oil). The additives are organized into micelles, in which paclitaxel becomes isolated, increasing the length of the retention time in the vascular network, and therefore prolonging the therapeutic effect of the drug. Unfortunately, these additives are very toxic and have many side effects.

Important progress has been made by radically changing the formulation of Taxol® and by replacing the additives present with albumin. The paclitaxel is then administered in the form of a formulation containing spherical paclitaxel NPs coated in albumin, called nab-paclitaxel (Abraxane®) (Figure 7.19).

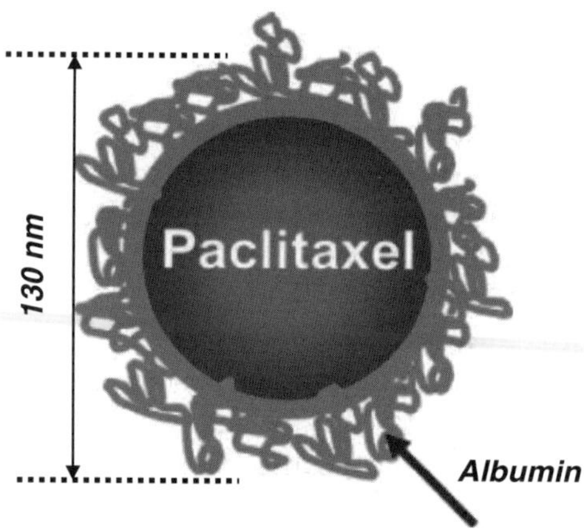

Figure 7.19. *Schematic representation of an Abraxane® NP (adapted from Desai [DES 12])*

16 Paclitaxel is the active compound of a natural product that is extracted from the bark of some Taxus trees (discovered in the 1960s) and that has the property of inhibiting cell division. It has proven to be a very efficient drug in the treatment of certain cancers (of breast, lungs). Taxol® was developed in France by the chemist Pierre Potier at the Institute of Natural Substances in Gif in the 1980s, by using another natural extract as a synthetic starting product, obtained in large quantities from the needles of different Taxus plants.

The pharmacodynamic properties of these NPs are at this stage much more in line with the required criteria for the targeted administration of the drug. First, the removal of Cremophor and the solvents considerably decreases the toxicity of the product, which means the prior administration of other drugs is no longer necessary; moreover, the hypersensitivity secondary reactions caused by Cremophor are no longer an issue. The size of the NPs, approximately 130 nm, is well adapted to the phenomenon of extravasation, and, furthermore, the presence of albumin on the surface of the paclitaxel is a targeting agent for the cancer cells. A particular protein (SPARC), which binds to albumin, is indeed secreted by the cancer cells, which considerably increases the selectivity of the treatment [IGL 09].

This new drug, which presents numerous advantages over Taxol®, could be considered to be the first commercial product to come out of the field of nanomedicine. It is completely stable as a solution and a lot less toxic than Taxol®, allowing higher dosage injections of paclitaxel, without increasing the hypersensitivity reactions.

7.3.3.2. *New paths in development*

The main problem that nanomedicine has to overcome, and one that conditions the optimization of a therapeutic treatment, is the issue of releasing the active principles of a drug at the site of the cellular lesion, and at that site only.

The ambitious goal is close to being reached. Already a few examples, still at the experimental stage, show that using a particular chemical engineering of NPs "loaded" with drugs could increase their therapeutic efficiency, while the migration of the active principles toward healthy tissue is kept to a minimum.

The use of degradable gelatin NPs, described in section 7.3.2.4, could be the solution. However, to our knowledge, they have not yet been put into use clinically.

Other solutions have been suggested. Among the more recent are the squalenoylation of gemcitabine, performed by the team headed by Professor Couvreur at the Faculty of Pharmacy of Châtenet-Malabry (France), the encapsulation of small interfering RNAs (siRNAs)[17] inside hybrid NPs (lipid liposomes coated in a particle of mesoporous silica), performed by the team headed by Professor Brinker in collaboration with several other American teams, in the Sandia Laboratories and several other research departments in the University of New Mexico.

7.3.3.2.1. Squalenoyl-gemcitabine

Gemcitabine is a nucleosidic analog of deoxycytidine, a compound that is highly analogous to the nucleotides that make up DNA (Box 3.3). It is a drug that is often used in the treatment of different cancers (of colon, lungs, pancreas, breasts, bladder, prostate, ovaries). Its therapeutic action is limited due to it being quickly degraded after intravenous injection, which requires inoculation in strong doses, which increases the toxicity of the drug.

These effects could be removed by isolating the gemcitabine from the plasmatic medium. This is done by binding a squalene (SQ), which completely modifies the physicochemical properties of gemcitabine, and in an aqueous medium the new molecules (SQ-gem) assemble themselves into nanostructures of cylindrical particles, in which the gemcitabine is dispersed, and becomes isolated from the blood plasma[18].

The key to the process is that the bond created between the gemcitabine and the SQ be an amide group that is easily broken by certain enzymes secreted by the tumor, which releases the active principles into the very heart of the cancer cells, and minimizes its dissemination into healthy tissue (Figure 7.20).

The clinical results for SQ-gem are indeed much better than those obtained with gemcitabine on its own, allowing the use of much lower doses of SQ-gem than with gemcitabine (Figure 7.21).

---

17 siRNAs are fragments of RNA that can silence certain genes when they enter a cell. RNAs are macromolecules comparable to DNA. Structurally, they differ from DNA in that the deoxyribose sugar is replaced by ribose sugars (ribose differ from deoxyribose in the presence of an OH group instead of a hydrogen atom on carbon 2), and the thymine is replaced by uracil (a base that has the same adenine-pairing properties as thymine). RNA can be seen as a copy of DNA. It is produced in the nucleus of cells with the help of enzymes (RNA-polymerase). It has an important role in the production of the proteins (messenger RNA), and can also be used as scaffolding in certain catalytic functions.

18 Squalene ($C_{30}H_{50}$) is a natural compound found in plants, animals and man, and especially in shark liver oil (hence the name). Commercially, it is produced by the extraction of fish oil. It is used as an adjuvant in certain vaccines.

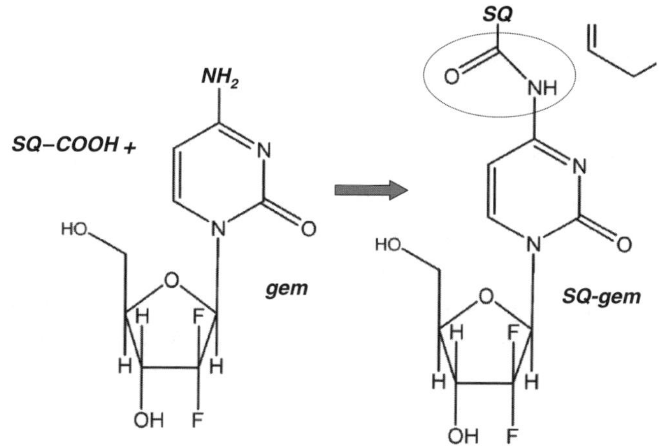

Figure 7.20. *The gemcitabine (gem) is functionalized by a derivative of SQ (SQ–COOH or $C_{26}H_{43}$–COOH), which binds to the gemcitabine by forming a NHCO bond (peptide bond) and giving up the SQ-gem. SQ comprises a long hydrophobic hydrocarbon chain that favors the formation of cylindrical NPs that contain a hydrophilic cavity in which the gemcitabine is confined. The NHCO bond is destroyed by the proteases (enzymes) that are produced by the tumor, which frees the gemcitabine into the heart of the tumor (adapted from Reddy et al. [RED 09])*

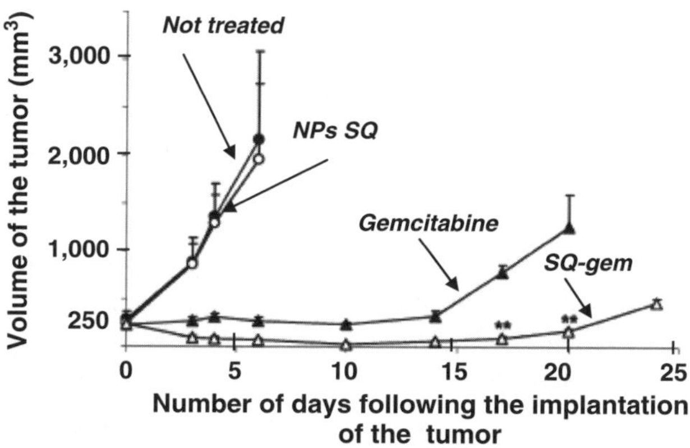

Figure 7.21. *Comparison of antitumoral activity between SQ-gem and gemcitabine NPs. The tumoral activity is estimated by measuring the increase of the volume of the tumor as a function of the number of days (strain of leukemic cancerous cells L1210, inoculated under the skin of a mouse). The non-treated tumor and the one in which SQ NPs were injected are placebos (adapted from Reddy et al. [RED 09])*

However, as the formulation with SQ is efficient, certain proteases (cathepsins B and D) must be present in the interstitial lysosome to be able to cut the NHCO peptide bond, which is necessary to release the gemcitabine.

7.3.3.2.2. Hybrid silica-lipid bilayer NPs

Gene therapy, which consists of introducing genes in a cell or diseased tissue to treat a pathology, is an old concept, which first appeared in the 1960s and developed considerably in the 1990s, notably with work by A. Fire and C.C. Mello (2006 Nobel Prize in Physiology or Medicine), which showed that certain genes can be silenced by introducing fragments of RNA (siRNA) in a cell. This discovery reactivated research in gene therapy, and currently a lot of hope is placed in treatments based on siRNAs able to interfere with RNA, thereby inhibiting certain genes responsible for a variety of diseases, especially cancers [LAU 08].

Their use as medication requires, as before, formulations that protect them from plasmatic breakdown and allows them to circulate long enough in the veins for them to be selectively directed toward the diseased tissue, and, after their internalization into the cell, for the active component to be released into the cellular cytosol and to intervene in a process called "gene silencing" (Figure 7.22).

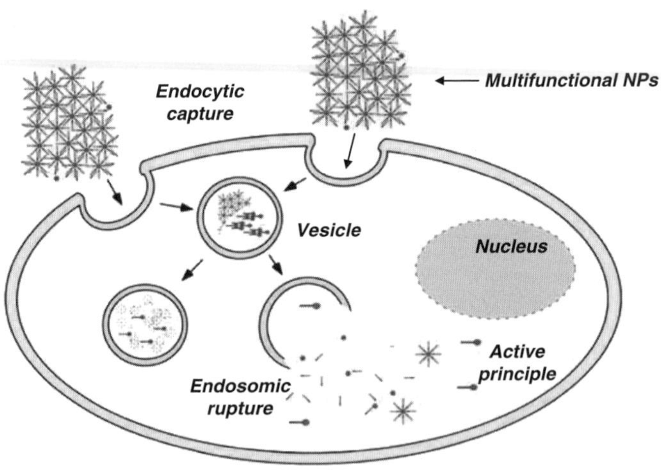

Figure 7.22. *Simplified diagram describing the penetration into a cell of the NPs due to the presence on the surface of the NPs of peptides that induce both the penetration of the NPs through the cellular membrane by endocytosis with the formation of a vesicle, and the endosomic rupture of the vesicle, which releases the active principle into the cytoplasm of the cell (adapted from Laufer [LAU 08])*

These different steps have recently been completed by making multifunctional hybrid NPs, which, after internalization inside the cell, are able to release siRNAs into the cellular cytosol [ASH 12].

The technique described consists of preparing mesoporous silica NPs (average diameter 165 nm, pores of 30 and 10 nm) that are "gorged" with the active principle by immersion of the NPs in a solution of siRNA.

These NPS are then fused with liposomes, functionalized with PEG, to confer to them the stealth required to avoid capture by the different macrophages of the immune system.

Functional peptides are also added at the surface of the lipid bilayer, one to help the NP target the membrane receptors characteristic of the disease, the other to insure the endosomic rupture of vesicle membrane (formed after the process of endocytosis) takes place (Figure 7.23).

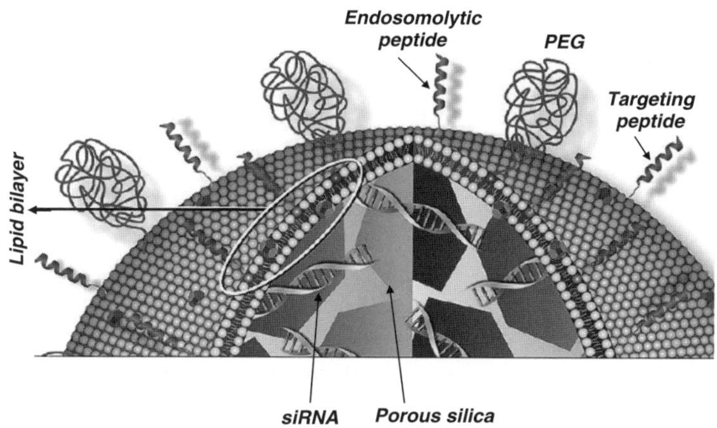

Figure 7.23. *Schematic exploded view representation of a hybrid NP (lipid bilayer-porous silica core) containing the siRNA in the silica pores. The PEG and the different peptides (targeting and endosomolytic) are bound to the lipid membrane as a result of the fusion of the liposomes with the silica NPs (adapted from Ashley et al. [ASH 12])*

The experiment was performed using a strain of hepatic cancer cells (Hep3B) and a mix of siRNAs chosen for their action on different proteins (cyclins) involved in the process of the production of cancer cells.

It is a sophisticated technique, which requires profound knowledge of the mechanisms involved in the function of different proteins, as well as perfect mastery of the techniques used to bind various molecules to the surface of bilayers. The liposomes that cover the silica are themselves the result of the combination of a mixture of natural surfactants (phospholipids and cholesterol) that guarantee good stability of the hybrid NPs in the blood, with hardly any loss of the active principle.

The results obtained are promising and clearly show that higher treatment efficiency is obtained for much lower doses of medication. Most importantly, it is shown that the storage capacity of the siRNA molecules is 10–1,000 times greater than that of liposomic particles of similar size; their stability in a neutral medium is also greatly improved.

The authors show, in the case of a hepatocellular carcinoma (Hep3B), that the different proteins involved in the tumor's growth process (cyclins A2, B1, D1 and E) are "silenced" by very weak concentrations of siRNA (0.090–0.37 nM), while with liposomic NPs, concentrations of 5–500 nM (about 1,000–10,000 times stronger) are required to induce the same effects [ASH 12].

This is huge progress when toxicity and efficiency, are considered. The results obtained, especially in the decrease in toxicity, will likely rapidly lead to the clinical phase of trials, which will determine the efficiency and safety of the new treatment.

7.3.4. *Problems related to the toxicity of nanomaterials*

Progress made in medical imaging as well as in nanomedicine is largely due to the use of various nanomaterials that are not without effect on living organism, and most importantly on human bodies. This has stirred a societal debate; however, no clear conclusion has been reached about the potential dangers of their dissemination or therapeutical use.

7.3.4.1. *A few general considerations*

The toxicity of materials has, until recently, been perfectly established in terms of chemical composition, and the European regulations REACH

(Registration, Evaluation, Authorization of Chemicals)[19] are perfectly clear on this point. REACH imposes a certain number of rules on companies that have contributed to the evaluation of the risk caused by the toxicity of certain chemical products, and to the definition of the conditions of use, until replacement with other non-toxic products.

The economic impact of these regulations is important, as it has caused the removal of certain products from the market and their replacement by other products, changes in composition, and a subsequent increase in production costs, which have important consequences on competition between companies.

The rise of nanomaterials is set to disturb previous evaluation systems, and already a number of recommendations have appeared to incite users of nanomaterials to take into consideration new toxicity criteria specific to this new class of products.

Because of their minute size, these NPs have a tendency to enter cells and start a cytotoxic process, which results in a disturbance of the physiological functions of the organism and a more or less severe alteration of the structure of the organs and the tissues. Generally, it is measured by the resulting effects in terms of mortality, by determining the threshold NPs concentration that leads to the death of a fraction of a particular type of biological cells[20].

It is given that among the factors that cause toxicity, the size, the shape of the particles, the chemical nature of the surface and its electrical charge, the composition and the stability of the NP are determining a whole range of criteria that are not taken into account in the current REACH regulations.

---

19 REACH is a set of regulations established within the European union, voted by the European parliament in 2006, that since 2008 has imposed on the chemical industry (importers, distributors and users of chemical products) of the European Union the registering of products and the evaluation of the toxicity and the risks involved in their handling and use (approximately 30,000 products), resulting in some restrictions on use. Taking into consideration the disparity of the regulations in other countries, another system of identification of the labeling of dangerous products called GHS (Globally Harmonized Stem of Classification and Labeling of Chemicals) was also introduced in 2008 by the United Nations, which caused the Europeans in 2010 to include the CLP (Classification, Labeling and Packaging) rules in REACH to harmonize with GHS.

20 The LD50 (lethal dose) is often used; it corresponds to the quantity of product necessary to kill 50% of the tested population.

This evaluation of toxicity, most often done *in vitro*, does not always represent behavior *in vivo*; it can also depend on the mode of administration of the NPs, which could be by inhalation, by mouth, intravenous, intradermic, which can all change the toxicity.

Taking into consideration the large number of factors involved, it could be supposed that the problem of evaluation toxicity is a complex one.

In the last decade, many studies concerning the toxicity of nanomaterials have been performed, often with the help of public health organizations, and there is currently a lot of data that can shed some light on the mechanisms of action of nanomaterials on living organisms. Some excellent reviews were published in 2011 by Soenen *et al.* [SOE 11] and Yildirimer *et al.* [YIL 11] that give a large overview of the works done up to now.

Essentially, evaluations of toxicity are done in the case of nanomaterials used therapeutically or in medical imaging (gold NPs, QDs, ferromagnetic NPs, silica NPs and polymer-based or phospholipid-based liposomes) and in the case of a few nanomaterials that are subject to important commercial applications, for example the new use of TiO_2, ZnO NPs in sunscreen, silver NPs with biocide properties, used in a number of sterilization processes.

From this important collection of information, which unfortunately is hard to correlate together due to the diversity of experimental protocols and targets, the conclusion is made that even some supposedly non-toxic NPs, already authorized clinically (gold NPs, ferromagnetic iron oxide NPs, silica NPs) can be toxic when used in certain conditions.

Thus, interestingly, gold NPs inoculated in the body of an albino mouse (BALB/C) are actually toxic if the particle size is between 8 and 37 nm, and non-toxic with particles of size 3, 5, 50 and 100 nm [CHE 09]. This toxicity, observed after injection of gold NPs over a period of one week (a ratio of 8 mg of gold particles per kilogram of living organism), results in the death of the mice 21 days later and in important changes of the hepatic, pulmonary and splenic cells.

These observations show that toxicity induced by nanomaterials is still poorly understood, and that total understanding of the mechanisms of NP-induced toxicity is far from obtained. This also contradicts a quotation from Paracelsus (16th Century) about chemical compounds that states, “All things are poison, and nothing is without poison; only the dose permits something not to be poisonous", to which might also be added "and size".

Such examples are not unique, and other peculiarities have also been observed with silica and iron oxide particles, which are normally considered non-toxic.

7.3.4.2. *The multiple causes of nanomaterial-induced toxicity*

Generally, as soon as a cell is exposed to a stress as the result of the presence of a "foreign" body or a pathogen, ROS are generated in excessive amounts in the surrounding environment. The species can destroy the functional proteins responsible for the health balance of the cell, react with the membrane phospholipids and lead to important cellular lesions causing inflammation and possibly even the death of the cell.

Such effects have been observed with Cd-based QDs, and also with inorganic NPs made of non-toxic metals like gold and ferric oxides. However, in the case of QDs that contain heavy metals like cadmium or selenium (CdS, SeS), toxicity could also be due to the presence in the cell of a small quantity of Cd^{2+} ions that are the result of the partial breakdown of the CdS NPs. It is also the reason for which the QDs are usually protected by a non-toxic inorganic layer (ZnS type), which is functionalized by a variety of ligands destined to minimize the process of dissolution of the Cd metal.

The breakdown of the NP caused by variations of the pH (acidity) during the process of endocytosis in the cells can also cause cytotoxicity. The pH is estimated to decrease locally, progressively from 7.4 (neutral) to 6, then to 4.5 (acidic), when the NP is internalized and is encapsulated in a lysosome (internal vesicle of the cell). This variation of pH, which corresponds to a concentration of H^+ ions a hundred times higher than when the pH was 7.4, is sufficient to partially dissolve the inorganic NPs (iron oxides, CdS Qds) and to release in the inside of the cell metallic ions of heavy metals that could be very toxic.

An example of such a dissolution is given by the curves of the variation in Figures 7.24(a) and (b) that show the quantities of ferric ions released by the iron oxide NPs as a function of the pH and of the time of immersion in the medium. At a slightly acidic pH (Figure 7.24(b)), the oxide iron NPs are broken down a lot quicker than at pH 7 and this breakdown is also less than when the NPs are coated in lipids.

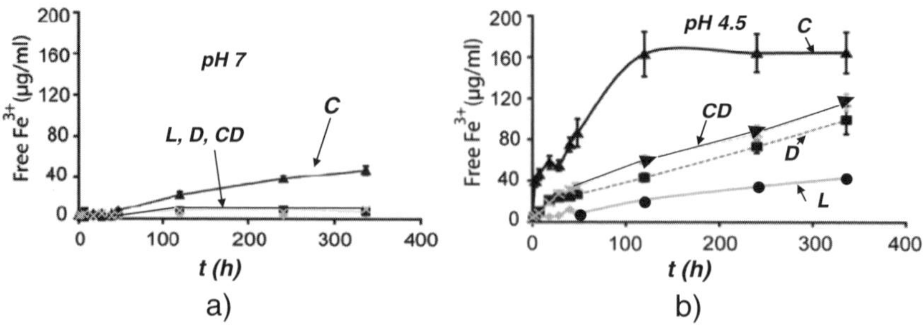

Figure 7.24. *Effect of pH on iron oxide NPs diversely functionalized by citrates (C), lipids (L), dextrane (D), carboxydextrane (CD). The curves show a small amount of dissolution of the iron oxide at pH 7 a), a lot of dissolution at pH 4.5 b), except for the NPs covered in lipids (L) (adapted from Soenen et al. [SOE 10])*

The morphological deterioration of the cells and of their cytoskeleton[21], a possible cause of cytotoxicity, can also take place after the internalization of the NP by the cell, a phenomenon that has been observed by several teams in the case of QDs and also with iron oxide and gold NPs [SOE 11].

Finally, the introduction of NPs into the cellular tissue, perceived as an intrusion of pathogenic elements, could also trigger a cascade of immune reactions, interact with proteins, various cellular receptors and block extremely complex biological processes. They are the subject of important research and this is also the reason why there is currently not enough information to get a clear idea about the causes of NP-induced toxicity.

7.3.4.3. *Recommendations for a better evaluation of NP toxicity*

The generalization of the use of nanomaterials in the multiple areas of industry has resulted in many studies being done on the sanitary and environmental impact that these materials might have. For example, in France, Afsset (French agency of sanitary security of work and the environment, or *Agence française de sécurité sanitaire de l'environnement et du travail*) led to evaluations on the risks for personnel in contact with nanomaterials. The CNRS, the INERIS (National Institutes for the Environment and Industrial Risk), the INRS (National Institute for Research and Security), the CEA (Commission for Atomic and Alternative Energies)

21 The cytoskeleton of a cell includes the biopolymers that are organized into filaments that can be found in the cytoplasm of a cell and give it its mechanical properties.

are also the main organizations involved in the study of risk and the precautions to be taken in the handling of nanomaterials.

Several international congresses on the theme of security are held periodically. The CEA, involved in problems of security, has organized the "Nano Safe" conferences every two years in Grenoble since 2008, which brings together the world's foremost experts[22] (the next "Nano Safe 2012" is to be held in November), during which all problems pertaining to security and toxicity related to nanomaterials (characterization, detection and surveillance, toxicology, nanomaterial life-cycle analysis, secure industrial production, risk analysis, standardization and regulations, etc.) are discussed.

A consensus is being established on the necessity of standardizing experimental protocol relative to the evaluation of the toxicity of nanomaterials. For this, a better definition of what constitutes a nanomaterial is a must.

It must take into consideration not only the chemical composition but also the nature of the chemical compounds (ligands) attached to the surface, the size and morphology of the NPs. The behavior of the NPs in a biological environment must also be mentioned, especially stability in a colloidal state (the presence or absence of particle aggregation). This would require knowledge of their surface potential (a property of the electrical charges) and their effective size in a solution (hydrodynamic diameter different from the dehydrated size).

The concentration of the NPs is also a parameter that should be better defined. Generally, the experiments are described in terms of the mass of the nanomaterials per unit of volume. This is perhaps not the most relevant parameter because the reactivity of NPs is more a function of their surface. On the contrary, knowledge of the number of NPs per unit of volume and of their developed surface are important parameters that should be taken into consideration in the future to establish possible correlations between nanomaterials and toxicity.

All these parameters are rarely combined in one single experiment. This explains the current difficulties to extrapolate simple cause and effect

---

22 The last conference Nano Safe 2010 is published in the *Journal of Physics: Conference Series*, vol. 304 (2011), available at http://iopscience.iop.org/1742-6596/304/1

relationships. Hopefully, in the future, strong recommendations will be expressed by different health and environmental organizations and will be accepted by the scientific community.

7.4. Conclusions and perspectives

Nanotechnology has become central both in biology and medicine. The increasing use of nanomaterials combined with the progress of physicochemical techniques for the analysis and detection of compounds at trace concentrations is currently revolutionizing the tools available for the diagnosis and treatment of the sick. The injection of QDs, gold NPs or even ferromagnetic NPs into a living organism, coupled with measures of luminescence, plasmon scattering and SERS, could result in an *in vivo* image that could help in getting an idea about the pathological state of the cellular tissue.

These new medical technologies, which mark an important break from the traditional methods of diagnosis and therapy, have been developed due to the help of chemists and physicists who synthesized "tailored" multifunctional nanomaterials and NPs with perfectly defined properties. These nanomaterials, functionalized to interact with the specific molecular targets of a disease, are then able to diagnose the zones affected by the disease and to transport and target a drug into the very heart of a diseased tissue.

A certain number of traditional drugs benefit from these new technologies and have already begun to improve treatments by minimizing their invasive nature. Although there is still a lot of progress to be made before being able to generalize the use of "nanovectorized" drugs in humans, it can be predicted that taking into account the constant increase in discoveries and innovations, in the next 10 years, the cure rate for diseases like cancer will have considerably improved.

Important progress must also be made in the area of understanding the mechanism of nanomaterial-induced toxicity. Too much variety in the experimental protocols used does not currently allow a good idea of what really causes this toxicity. The standardization of the experimentation is strongly recommended, and taking into consideration relevant factors such as the overall surface of the NPs as well as their stability in a biological medium should allow a better grasp in the future of the parameters responsible for their toxicity.

7.5. Bibliography

[AMA 11] AMATORE C., ARBAULT S., GUILLE M., LEMAITRE, F., "Artificial synapses and oxidative stress", *Actualité Chimique* , vol. 348–349, pp. 25–31, 2011.

[ASH 12] ASHLEY C.E., CARNES E.C., EPLER K.E., PADILLA D.P., PHILLIPS G.K., CASTILLO R.E., WILKINSON D.C., WILKINSON B.S., BURGARD C.A., KALINICH R.M., TOWNSON J.L., CHACKERIAN B., WILLMAN C.L., PEABODY D.S., WHARTON W., BRINKER C.J., "Delivery of small interfering RNA by peptide-targeted mesoporous silica nanoparticle-supported lipid bilayers", *ACS Nano*, vol. 6, pp. 2174–2188, 2011.

[CHE 09] CHEN Y.-S., HUNG Y.-C. LIAU I., HUANG G.S., "Assessment of the in vivo toxicity of gold nanoparticles", *Nanoscale Research Letters*, vol. 4, pp. 858–864, 2009.

[CHO 10] CHOI H.S., LIU W., LIU F., NASR K., MISRA P., BAWENDI M.G. FRANGIONI J.V., "Design considerations for tumor-targeted nanoparticles", *Nature Nanotechnology*, vol. 5, pp. 42–47, 2010.

[CUI 01] CUI Y., WEI Q., PARK H.K., LIEBER C.M., "Nanowire nanosensors for highly sensitive and selective detection of biological and chemical species", *Science*, vol. 293, pp. 1289–1292, 2001.

[DES 12] DESAI N., "Challenges in development of nanoparticle-based therapeutics", *The AAPS Journal*, vol. 14, pp. 282–295, 2012.

[ENG 10] ENGEL Y., ELNATHAN R., PEVZNER A., DAVIDI G., FLAXER E., PATOLSKY F., "Supersensitive detection of explosives by silicon nanowire arrays", *Angewandte Chemie International Edition*, vol. 49, pp. 6830–6835, 2010.

[GAR 07] GARCIA M., JEMAL A., WARD E.M., CENTER M.M., HAO Y., SIEGEL R.L., THUN M.J., *Global Cancer Facts & Figures 2007*, American Cancer Society, Atlanta, GA, 2007.

[IGL 09] IGLESIAS J., "*nab*-Paclitaxel (Abraxane®): an albumin-bound cytotoxic exploiting natural delivery mechanisms into tumor", *Breast Cancer Research: BCR*, vol. 11, suppl. 1, p. S21, 2009.

[JUN 09] JUN Y.-W., SHEIKHOLESLAMI S., HOSTETTER D.R., TAJON C., CRAIK C.S., ALIVISATOS A.P., "Continuous imaging of Plasmon rulers in live-cells reveals early-stage caspase-3 activation at the single molecule level", *Proceedings of the National Academy of Sciences of the United States of America*, vol. 106, pp. 17735–17740, 2009.

[KIR 10] KIRUI D.K., REY D.A., BATT C.A., "Gold hybrid nanoparticles for targeted phototherapy and cancer imaging", *Nanotechnology*, vol. 21, p. 105105 (10 p), 2010.

[KOO 11] KOO H., HUH M.S., RYU J.H., LEE D.-E., SUN I.-C., CHOI K., KIM K., KWON I.C., "Nanoprobes for biomedical imaging in living systems", *Nano Today*, vol. 6, pp. 204–220, 2011.

[LAU 08] LAUFER S.D., RESTLE T., "Peptide-mediated cellular delivery of oligonucleotide-based therapeutics in vitro: quantitative evaluation of overall efficacy employing easy to handle reporter systems", *Current Pharmaceutical Design*, vol. 14, pp. 3637–3655, 2008.

[LEE 08] LEE H., LEE K., KIM I.K., PARK T.G., "Synthesis, characterization, and in vivo diagnostic applications of hyaluronic acid immobilized gold nanoprobes", *Biomaterials*, vol. 29, pp. 4709–4718, 2008.

[MIS 09] MISRA N., MARTINEZ J.A., HUANG S.-C.J., WANG Y., STROEVE P., GRIGOROPOULOS C.P., NOY A., "Bioelectronic silicon nanowire devices using functional membrane proteins", *Proceedings of the National Academy of Sciences of the United States of America*, vol. 106, pp. 13780–13784, 2009.

[NIE 10] NIE S., "Understanding and overcoming major barriers in cancer nanomedicine", *Nanomedicine*, vol. 5, pp. 523–528, 2010.

[NOY 11] NOY A., "Bionanoelectronics", *Advance Materials*, Vol. 23, pp. 807–820, 2011.

[PAP 91] PAPAHADJOPOULOS D., ALLEN T.M., GABIZON A., MAYHEW E., MATTHAY K., HUANG S.K., LEE K.-D., WOODLE M.C., LASIC D.D., REDEMANN, C., MARTIN F.J., "Sterically stabilized liposomes: improvements in pharmacokinetics and antitumor therapeutic efficacy", *Proceedings of the National Academy of Sciences of the United States of America*, vol. 88, pp. 11460–11464, 1991, available at http://www.pnas.org/content/88/24/11460.full.pdf

[PHI 10] PHILLIPS M.A., GRAN M.L., PEPPAS N.A., "Targeted nanodelivery of drugs and diagnostics", *Nano Today*, vol. 5, pp. 143–159, 2010.

[POP 10] POPOVIC Z., LIU W., CHAUHAN V.P., LEE J., WONG C., GREYTAK A.B., INSIN N., NOCERA D.G., FUKUMURA D., JAIN R.K., BAWENDI M.G., "A nanoparticle size series for in vivo fluorescence imaging", *Angewandte Chemie International Edition*, vol. 49, pp. 8649–8652, 2010.

[RED 09] REDDY L.H., RENOIR J.-M., MARSAUD V., LEPETRE-MOUELHI S., DESMAËLE D., COUVREUR P., "Anticancer efficacy of Squalenoyl Gemcitabine nanomedicine on 60 human tumor cell panel and on experimental tumor", *Molecular Pharmaceutics*, vol. 6, pp. 1526–1535, 2009.

[SOE 10] SOENEN S.J., HIMMELREICH U., NUYTTEN N., PISANIC T.R., FERRARI A., DE CUYPER M., "Intracellular nanoparticle coating stability determines nanoparticle diagnostics efficacy and cell functionality", *Small*, vol. 6, pp. 2136–2145, 2010.

[SOE 11] SOENEN S.J., RIVERA-GIL P., MONTENEGRO J.-M., PARAK W.J., DE SMEDT S.C., BRAECKMANS K., "Cellular toxicity of inorganic nanoparticles: common aspects and guidelines for improved nanotoxicity evaluation", *Nano Today*, vol. 6, pp. 446–465, 2011.

[WHI 06] WHITESIDES G.M., "The origins and the future of microfluidics", *Nature*, vol. 442, pp. 368–373, 2006.

[WON 11] WONG C., STYLIANOPOULOS T., CUI J., MARTIN J., CHAUHAN V.P., JIANG W., POPOVI Z., JAIN R.K., BAWENDI M.G., FUKUMURA D., "Multistage nanoparticle delivery system for deep penetration into tumor tissue", *Proceedings of the National Academy of Sciences of the United States of America*, vol. 108, pp. 2426–2431, 2011.

[YIL 11] YILDIRIMER L., THANH N.T.K., LOIZIDOU M., SEIFALIAN A.M., "Toxicological considerations of clinically applicable Nanoparticles", *Nano Today*, vol. 6, pp. 585–607, 2011.

[ZHE 05] ZHENG G., PATOLSKY F., CUI Y., WANG W.U., LIEBER C.M., "Multiplexed electrical detection of cancer markers with nanowire sensor arrays", *Nature Biotechnology*, vol. 23, pp. 1294–1301, 2005.

Chapter 8

Nanorobotics and Nanomachines of the Future

The idea of molecular machines able to perform basic functions of movement, transport and elevation requires the construction at a molecular level of equivalent mechanisms, such as rotors, motors and other systems, for which a lot of knowledge and technique in chemical synthesis are needed.

Nature gives many examples of extremely sophisticated nanomachines, able to perform a number of biological functions in an aqueous environment. For example, ATP-synthases are minute rotating motors, contained within a mitochondrial membrane, which catalyze the conversion of adenosine diphosphate (ADP) into adenosine triphosphate (ATP), which has more energy than ADP. Other example includes kinesin, a protein that in the presence of ATP is able to travel along a microtubule (fibers that make up the cellular cytoskeleton), myosin, another protein that plays an important role in the mechanism of muscular tissue contraction, or even bacterial flagella that are equipped with a molecular motor that allows them to move in liquid.

The perfection of these biological systems dedicated to the accomplishment of precise biological functions can only amaze; this explains why active research is constantly going on designing artificial molecular machines that attempt to imitate them under the effect of a variety of stimuli (photochemical, electrical and chemical).

Synthetic molecular machines are clearly still a long way from the functional possibilities of natural biological systems. However, a large variety of chemical compounds have been thought up based on the two structural models of rotaxane and catenane, through the binary assembly of molecules in liquid.

With rotaxanes, one of the molecules slides on an axis by making linear movements, whereas with catenanes, which can be compared to two links in a chain, one of the molecules makes rotational movements around the other that is static.

These molecular movements can be set off by chemical, photochemical or electrical stimuli, opening the door for important applications in the domain of nanomedicine, especially with the conception of molecular valves able to deliver medication under the effect of external impulses.

The final objective is to get as close as possible to the function of natural molecular machines, which are made up of the complex assemblies of proteins. ATP-synthase and myosin are two typical examples.

8.1. Natural molecular machines

ATP-synthase and myosin are the two best-known examples of molecular machines. They can give an idea of the complexity of their functioning and of their efficiency in creating movement at a nanometric level.

8.1.1. *ATP-synthase*

This is a particular ATPase (enzyme responsible for the conversion of ADP into ATP) that is found in plants, bacteria and the mitochondrial membranes of eukaryote cells[1].

Inserted inside a mitochondrial membrane, it is sometimes presented as a molecule turbine that uses proton flux energy (created by a difference of concentrations between the outside and inside of the membrane) to activate a

1 Eukaryote cells (as opposed to prokaryote cells) contain a nucleus and mitochondria in which energy exchange takes place as a result of the ADP $\leftrightarrow$ ATP conversion.

molecular rotor, the energy of which is used to synthesize ATP from ADP and phosphate according to the reaction ADP + Phosphate (P_i) + $H^+ \rightarrow$ ATP (Figure 8.1).

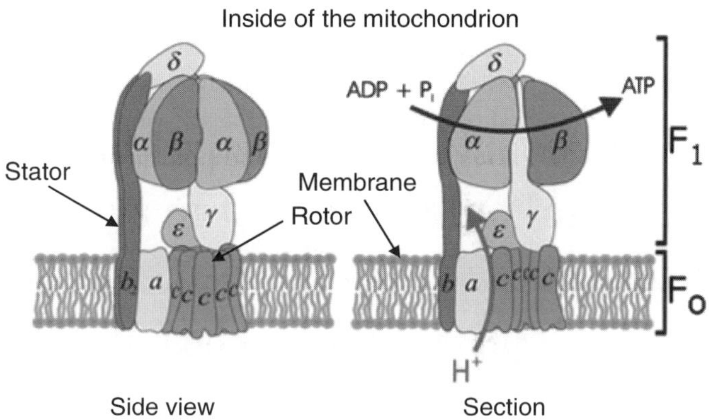

Figure 8.1. *Schematic representation of ATP-synthase, split into two parts F_0 and F_1 (F_1 inside the mitochondrion). The rotor **c** is prolonged by an axis γ surrounded by the proteins $\boldsymbol{\alpha}$ and $\boldsymbol{\beta}$. The H^+ protons diffuse from the inside of the cell toward the inside of the mitochondrion through two semi-channels located in $\boldsymbol{\alpha}$, one communicating with bottom and the other with top of the membrane. The binding of a proton on a protein site causes the deformation of the protein, creating a couple necessary for the rotation of the rotor. The proton is then released inside the mitochondrion after one full rotation of the rotor, and used to synthesize an ATP molecule (adapted from B.A. Feniouk, http://www.atpsynthase.info/Gallery.html)*

The catalytic region comprises nine protein subunits (3α, 3β, 1γ, 1δ, 1ε). The protons travel in channels located between the stator (a, b, δ) and the rotor (c, γ, ε). They cause the sequential rotation of the rotor, leading to the production of ATP, as a result of the action of the three active sites α, β, δ (Elston *et al.* [ELS 98], Stock *et al.* [STO 99], Vale and Milligan [VAL 00], Oster and Wang [OST 03]).

It is a very complex mechanism that relies on astonishing molecule organization. The discovery of the chemistry that converts ADP into ATP was done by P.D. Boyer (the USA) and J.E. Walker (GB), for which they were awarded the Nobel Prize in chemistry in 1997, shared with J.C. Skou (Denmark), for his work on the transport enzyme, Na+, K+-ATPase.

8.1.2. *Myosin: a linear protein nanomotor*

It is the energy released during the hydrolysis of ATP into ADP that induces the linear movement of myosin along the actin filaments.

Muscles are made up of fibers that themselves are a repetition of segments (sarcomeres), which are an assembly of intermingled actin and myosin filaments. It is inside each sarcomere that the phenomenon of muscle contraction takes place, corresponding to the slide of the actin filaments over the myosin filaments (Figure 8.2).

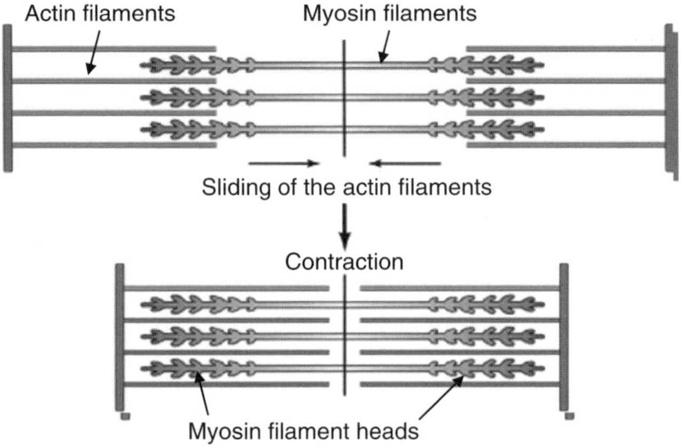

Figure 8.2. *Schematic representation of the contraction mechanism of a muscle fiber by the sliding of actin filaments over myosin filaments inside a sarcomere. The myosin filament heads correspond to the active part of the myosin that interacts with the actin filament on specific sites (adapted from a document of the National Center for Biotechnology Information (NCBI), http://www.ncbi.nlm.nih.gov/books/NBK9961/)*

The mechanism is complex and several physicochemical stimuli are at the origin of the contraction phenomenon. The arrival of calcium ions caused by a nervous influx first results in a conformational change in the actin filaments, allowing the myosin heads (a sort of protuberance) to bind to the active sites of the actin. The myosin heads then undergo a conformational change that results in the actin filament sliding a few nanometers toward the inside of the sarcomere, causing its contraction.

It is the repetition of this process, energetically performed by the hydrolysis of ATP, that produces the forward motion of the actin filament toward the center of the sarcomere.

These two examples show how nature is able to create high-precision molecular machines, of great complexity.

Using natural biological systems as inspiration, important research has developed toward the synthesis of artificial molecular machines that recognize information, store and transmit it and, under the effect of an energetic stimulus, perform linear or rotational controlled directional movements.

8.2. Artificial molecular machines

A large part of research has thus been focused on the conception of systems in a solution or on surfaces, and relies on principles that apply to supramolecular chemistry, where the formation of molecular systems is determined by low-energy interaction forces.

Generally, these artificial molecular machines comprise two molecular components, one of which is static in relation to the other, and likely to occupy an equilibrium position under the effect of an external impulse that can be chemical, electrochemical or photochemical. These equilibrium positions correspond to the different minimal energy levels of the process.

Two sorts of molecular construction have been designed.

The first is characterized by a mobile component connected to a rigid frame by two strong chemical bonds (covalent bonds). Axially symmetrical, these bonds allow free rotation of the mobile component, which can thus pivot and lead to different shapes corresponding to favored conformations. Such systems have functions comparable to those of rotors, helices, gears, tourniquets, etc. (Figure 8.3).

The second type of molecular machine is characterized by two components, with no strong chemical bond between them, and corresponds to interlaced chemical structures, such as rotaxanes and catenanes. These are by far the most studied structures; the rotaxanes have already found promising application in the domain of the vectorization of drugs.

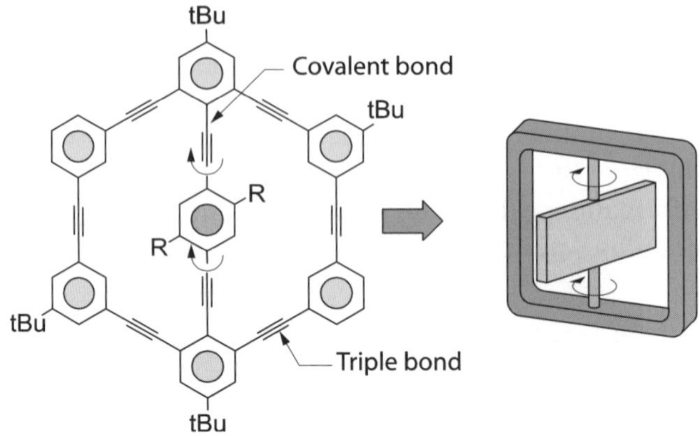

Figure 8.3. *Example of a molecular machine comparable to a tourniquet. The process's mobile section (rotor) is inserted inside a cyclical component (macrocycle) and joined to this cycle by two covalent bonds around which a free rotation takes place. The triple bonds guarantee the rigidity and the linearity of the sequences between the carbon atoms. The smaller the R groups carried by the central phenyl nucleus, the faster the rotation. It can be blocked if the R group is very big (adapted from Balzani et al. [BAL 00]*

8.2.1. *Artificial molecular machines in solution*

Two principal types of molecular construction have been used to make molecular "shuttles", systems in which a molecule can be moved to and from different sites in a host molecule. These are usually two structural models corresponding to structures of chemical compounds known as "rotaxane" and "catenane".

A rotaxane can be defined as a rigid linear molecule (axis in the shape of a dumbbell) over which slides another molecule, which can perform a back-and-forth movement, or a translation movement (Figure 8.4). This movement is confined to the axis thanks to the presence at the two extremities of large blocking groups ("stoppers").

A catenane is a system that corresponds to two interlocked molecules, one of them being able to change position as a result of a rotation (Figure 8.5).

The creation of these chemical systems uses concepts of supramolecularity and was mainly developed by J.F. Stoddart, considered to be the initiator of such research (Birmingham, UK, and University of

California, Los Angeles), V. Balzani in Italy (University of Bologna) and J.P. Sauvage in France (University of Strasbourg).

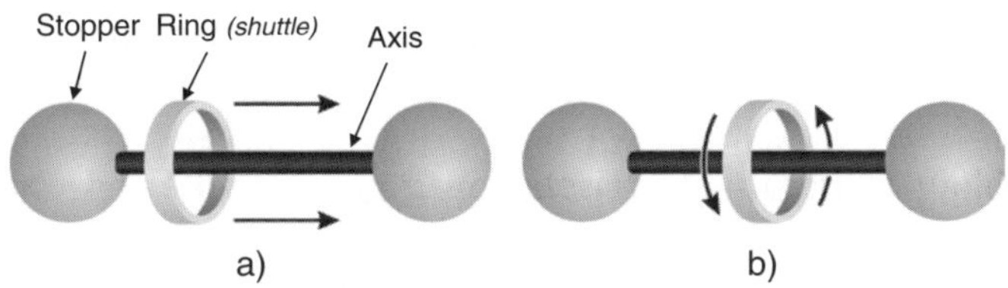

Figure 8.4. *Schematic representation of a rotaxane. It is composed of a rigid molecular axis and of a ring-shaped molecule (shuttle) that can slide along the axis by either a) translation or b) rotation. The ring remains confined to the axis due to the presence at the two extremities of large blocking molecules (stoppers) that stop it from leaving the axis (adapted from Silvi et al. [SIL 09])*

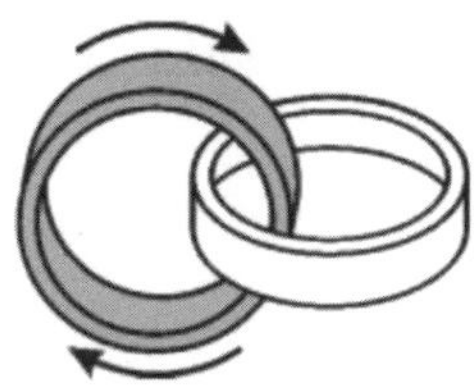

Figure 8.5. *Schematic representation of a catenane. The two molecular systems are interlocked with each other. One can be acted on to make it pivot (adapted from Balzani et al. [BAL 06])*

8.2.1.1. *Rotaxanes (translational molecular shuttles)*

The controlled movement of a molecule (ring-shaped), sliding along an axis, was first performed by J.F. Stoddart's team in the case of a rotaxane. The rotaxane's axis contains two separate sites ("stations"), one of which can be stimulated and can position the molecular ring on one of the two sites.

By judiciously adapting the chemical structure of rotaxane, the movement of the shuttle with a programmed stop of the ring at one of the two "stations" can be reached through the action of different external stimuli such as a variation in acidity, a change in the oxidative state of a redox group, or even the photochemical excitation of a photosensitive group.

8.2.1.1.1 Acidic stimulus-powered shuttles

The rotaxane is chosen so as to have two separate nitrogen sites. One is permanently positively charged (bipyridinium group, $BiPy^{2+}$), the other is an amino neutral group (NH) (no charge in a neutral solution) that after protonation in an acidic solution acquires a positive charge (NH_2^+)[2].

The ring (shuttle) is a macrocycle known as a crown-ether, characterized by the presence of several sequences of ether-oxide $-OCH_2CH_2O-$. These oxygen atoms are drawn toward the positively charged sites, resulting in the preferential positioning of the shuttle on the NH^{2+} and $BiPy^{2+}$ sites (Figure 8.6).

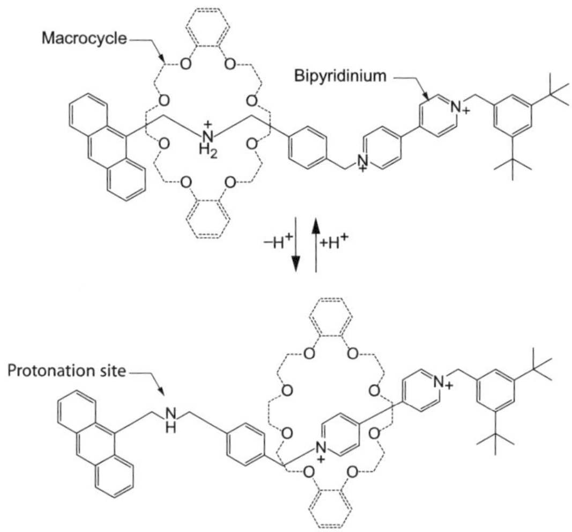

Figure 8.6. *Example of a molecular shuttle comprising a macrocycle as shuttle and an axis with two "stations" (bipyridinium^{2+} and amino groups $NH \leftrightarrow NH_2^+$). In an acidic medium, the NH group acquires a H^+ proton and becomes positively charged, the macrocycle interacts preferentially with NH_2^+ and therefore positions itself at this level. In a neutral medium, NH_2^+ loses the H^+ proton, the interaction with the macrocycle disappears and, in this case, it is the interaction with the bipyridinium^{2+} group (weaker than with NH_2^+) that dominates and binds the position of the shuttle to the bipyridinium^{2+} station (adapted from Ashton et al. [ASH 98])*

The macrocycle has a much higher affinity for the ammonium group than the $BiPy^{2+}$ group. In an acidic medium, the shuttle is therefore preferentially positioned on the NH_2^+ group and, in a neutral medium, due to the fact that NH_2^+ returns to the non-charged NH, localization on the $BiPy^{2+}$ group, which is the only one that remains positively charged.

2 Each positive electric charge is equal in absolute terms to an electron's negative charge.

The system has dimensions in the order of 2–3 nm; it functions reversibly at room temperature when the medium is successively acidified and then neutralized.

Many other systems have been thought up following this design, often with the same functional units containing a macrocycle with carbon–oxygen bonds as a shuttle and bipyridinium and ammonium groups as positioning stations. With some structures, the effects of contraction and dilatation of the chains have been observed, systems that are then considered to be analogous to molecular muscles (Liu *et al.* [LIU 05]).

8.2.1.1.2. Redox stimulus-powered shuttle

The principle is the same but, unlike the previous case, the "sliding cycle" (shuttle) is now made up of a positively charged cyclical compound, containing two bipyridinium ($BiPy^{4+}$) nuclei that are perfectly stable in the experimental conditions, which correspond to four positive electrical charges.

The main chain comprises two neutral sites: the tetrathiafulvalene (TTF) and the 1,5-dioxynaphtalene (DNP). Each site presents a strong affinity for the cycle $BiPy^{4+}$, although the TTF is stronger, which means that in the neutral state the shuttle is preferentially positioned on the TTF station (Figures 8.7(a) and (b)).

By adding an oxidant, the TTF is oxidized and gains two positive charges (TTF^{2+}), which results in the $BiPy^{4+}$ cycle being pushed away toward the second non-charged DNP station (Figure 8.7(c)).

The system has been tested in the case of molecules bound to silica (SiO_2) particles, through bonds between OCH_3 terminals and SiOH functions present at the surface of the silica.

It must be noted that the back-and-forth function of the shuttle requires every time the introduction of different reagents: an oxidant to convert TTF to TTF^{2+} and to move the shuttle from the TTF station toward the DNP station, a reducing agent to bring it back to the TTF. This is clearly quite inconvenient, but can be overcome by using physical stimuli and, most importantly, light energy, which does not require any kind of supplementary material. The function of the shuttle is then considered to be autonomous.

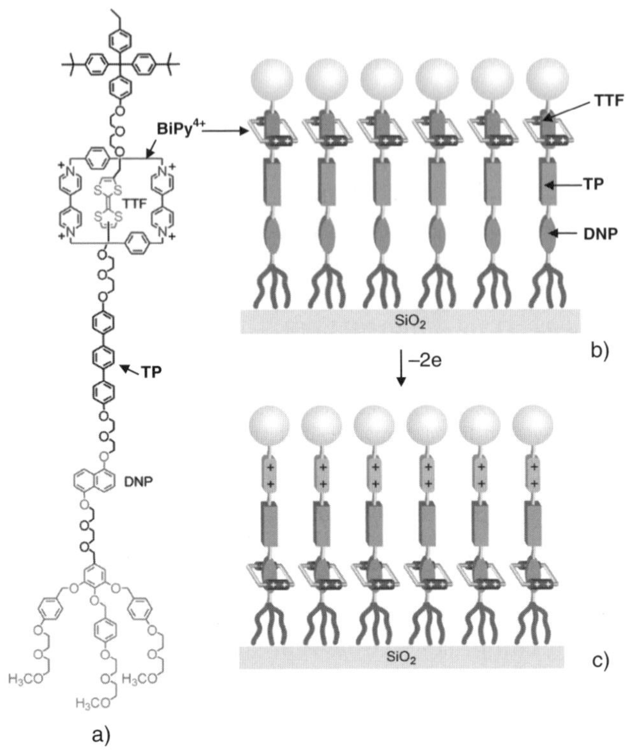

Figure 8.7. *(a) Molecular representation of the "BiPy^{4+} shuttle - molecular chain with TTF and DNP stations". The triphenyl (TP) causes the principle chain to become rigid and plays no part in the positioning of the shuttle; (b) the shuttle is positioned on the TTF station when it is in its neutral state; (c) the shuttle migrates toward the DNP when the TTF is oxidized to TTF^{2+} (adapted from Silvi et al. [SIL 09]). For a color version of this figure see www.iste.co.uk/lacaze/nano.zip*

8.2.1.1.3. Photochemical stimulus-powered shuttle

The principle is the same as before, except in the mode of excitation.

The absorption of light by a photosensitive chemical group (porphyrin P), incorporated in the axial chain, causes the movement of the shuttle, which, like previously, is made up of two BiPy^{2+} groups.

Such a device was also designed and synthesized by Stoddart's team. The system is multifunctional and comprises two separate parts: one region containing the porphyrin in which the light is absorbed and the fullerene (electron acceptor) where the charge transfer takes place. The other region, in which the shuttle moves, is controlled by a reduction–oxidation reaction between the TTF and the porphyrin.

Starting from a neutral state, where the TTF station is surrounded by the "two $BiPy^{2+}$" shuttle, a succession of four separate events must take place to induce the migration of the shuttle toward the DNP station:

– absorption of a photon by porphyrin,

– charge separation and transfer of an electron from the "excited" state of the porphyrin toward an electron acceptor (fullerene C60),

– oxidation of the TTF by the porphyrin (P) and conversion of TTF into TTF^+,

– movement of the shuttle toward the DNP station (Figure 8.8).

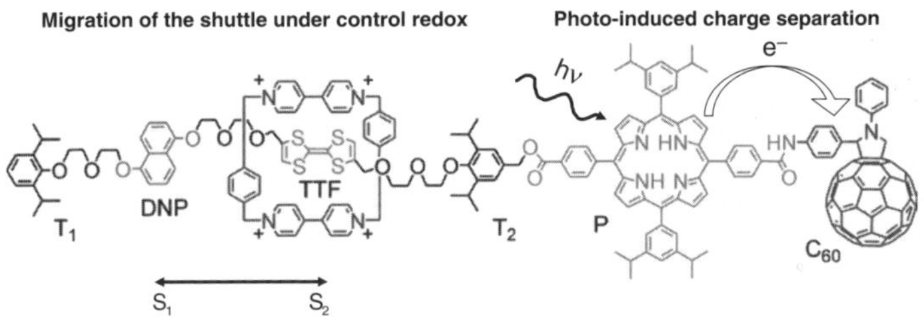

Figure 8.8. *Photochemically stimulated shuttle. The absorption of a photon (hv) by porphyrin (P) causes a charge separation between P and fullerene (C60), which corresponds to the transfer of an electron from P to C60. The porphyrin gains a positive charge (oxidized state) and can then go on to exchange this charge with the TTF group (station S_2), which is easily oxidized. The positive TTF pushes the shuttle toward the neutral DNP (station S_1) (adapted from Saha et al. [SAH 07])*

The last step of reduction–oxidation corresponds to an exchange of electrons between the porphyrin and the TTF, and is surprising considering the distance between the two units shown in Figure 8.8. This implies that the entirety of the system is not linear like the drawing would suggest, but that in a solution it acquires a geometry that allows the TTF and P groups to come in contact with each other to exchange an electron, an effect that takes place frequently in the case of biological systems involving proteins.

8.2.1.2. *Catenanes*

These are two macrocycles interlocked with each other. One of the cycles contains "acceptor" sites (chemical groupings carrying positive charges), while the other possesses "donor" sites (chemical groups rich with electrons). The commutation between the two "stations" is obtained as an acceptor site gets closer to a donor site, which thermodynamically corresponds to a state of low energy and, therefore, to a stable preferential conformation of the catenane.

A large variety of systems has been described and, in most cases, the movements produced are rotations of one cycle in relation to the other, caused by external stimuli, which like previously, can be chemical, electrochemical or photochemical.

The mechanics are simple and correspond in essence to those described in Figure 8.9.

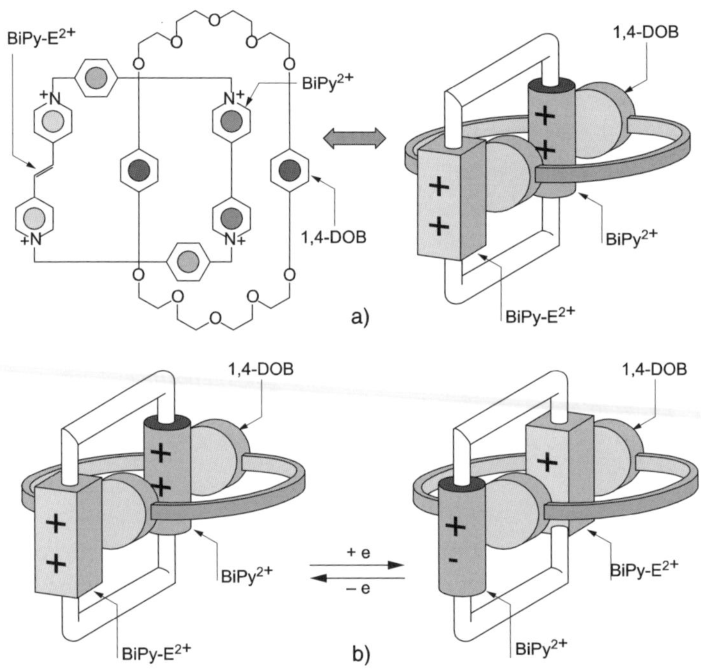

Figure 8.9. *Commutation mechanism of a tetracationic cyclophane (4+) interlocked with an ether-crown macrocycle containing two electron-rich 1,4-DOB groups: a) Schematic representation of the catenane ($BiPy^{2+}$ is represented by a cylinder, BiPy-E^{2+} by a rectangular parallelepiped); b) commutation set off by the electrochemical reduction (addition of an electron e) of $BiPy^{2+}$ to $BiPy^{1+}$, which causes the permutation of the positions of $BiPy^{2+}$ and BiPy-E^{2+} inside the crown-ether. The commutation takes place at room temperature in a solvent (acetonitrile) (adapted from Balzani et al. [BAL 00])*

The catenane contains a tetracationic cyclophane with two separate positively charge sites (a $BiPy^{2+}$ and a $bipyridinium^{2+}$ ethylene, or $BiPyE^{2+}$) and a neutral crown ether oxide symmetrically bound to two 1,4-dioxybenzene groups ($O–C_6H_4–O$, or 1,4–DOB), which are strongly attracted to the positive sites (due to the high electron density of the DOB groups).

In its most favorable initial configuration, the $BiPy^{2+}$ group is "sandwiched" between the two 1,4–DOB donor nuclei (Figure 8.9(a)). The electrochemical reduction of catenane (transfer of an electron toward $BiPy^{2+}$) converts $BiPy^{2+}$ into $BiPy^{1+}$ (loss of a positive charge) whose affinity toward the 1,4–DOB is less than that of ($BiPy\text{-}E^{2+}$), which always contains two positive charges.

The loss of the cationic charges results in the rotation of the $BiPy\text{-}E^{2+}$ group such as to obtain a configuration that is energetically more favorable, which comes down to permuting the position of $BiPy^{2+}$ and $BiPy\text{-}E^{2+}$ between the two 2,4–DOB groups (Figure 8.8(b)).

8.2.1.3. *Promising applications for diagnosis and therapy*

This prospective and fundamental research has for its goal to be able to integrate these molecular machines into artificial processes to make them carry out some of the cycles of chemical conversions normally done by natural molecular machines. Such objectives have already been achieved, for example with the conception of an artificial system, inserted in the membrane of a liposome, which can reproduce the function of a proton, required to activate ATP-synthase (Steinberg-Yfrach *et al.* [STE 98]).

Other more recent applications show that the previously described molecular shuttles can be used as molecular valves, and in this way considerably improve the delivery strategy of drugs into the organism, both for therapy and diagnosis.

A technique that is becoming increasingly widespread consists of using nanoparticles of porous silica of 100 nm with pores of 2 nm in diameter, in which drugs are inserted in high concentrations.

This type of vector is increasingly used because of its non-toxicity, its robustness and its easy uptake by living cells.

Moreover, the possibility of easily binding different reagents on their surface using organo-siloxane (–SiOR) functions is also an advantage that makes these devices very attractive for applications in diagnosis or therapy.

The challenge is to bind to the inside of the pores and on their walls functional compounds like molecular machines, designed to release the drugs into the pores under the effect of an external impulse.

Such devices represent additional progress over the systems based on "passive" silica NPs that continuously and spontaneously deliver the drugs initially stored in the pores (see Chapter 7, section 7.3.3.2.2).

Different "active" systems have thus been thought up by applying the previously described concepts. Depending on the case, the stimuli can be changes in acidity, reduction–oxidation reactions, thermic or photochemical impulses, or even specific interaction with biological molecules.

The use of *cis–trans* isomerization of azo compounds (rigid compounds with double bonds between two nitrogen atoms) bound to the walls of silica NP walls is particularly ingenious and has already been tested for use on cancer cells (Lu *et al.* [LU 08]).

The opening of silica channels is in this way controlled by a *trans-cis* isomerization reaction caused by the absorption of light. The process is well known to chemists and corresponds to an inversion of the position of a chemical group bound to a nitrogen atom, as in Figure 8.9.

It is this model that has been used to perform the controlled obturation and closing of the silica channels, which had been previously "gorged" with therapeutic molecules (Figure 8.10).

Such possibilities are extremely promising and are progressive in the administration of drugs in limiting the invasive effects of the drugs in the organism. The setup was tested *in vivo* on cancer cells, and it was shown that molecules with anti-cancer properties stored in the pores of silica NPs could be delivered on command and penetrate inside cells, after irradiation of the NPs.

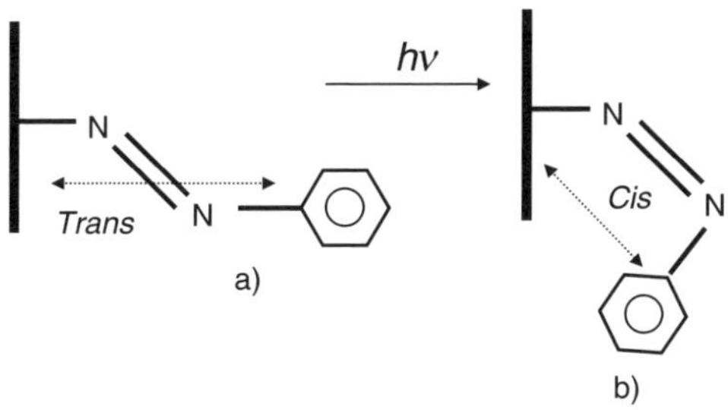

Figure 8.10. *Photochemical trans → cis isomerization of a nitrogen derivative (azo compound) bound to a solid wall: a) In its most stable configuration, the compound adopts the trans form; b) after irradiation (hv), the compound adopts the cis configuration, which involves the folding of the benzene nucleus toward the wall, and therefore freeing the space inside the silica channel. The compounds return to its initial configuration if heated or excited at a wavelength different from that corresponding to the trans → cis transition*

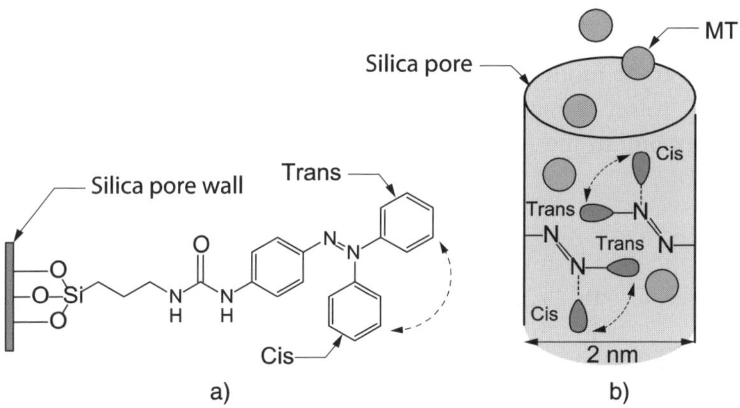

Figure 8.11. *Schematic representation of a molecular valve controlled by light irradiation: a) Azo derivative bound to the internal wall of a silica NP pore through siloxane (Si–O) bonds. The benzene nucleus in the trans position switches to the cis position after irradiation; b) relative disposition of the benzene nuclei inside a silica pore. The pore is obstructed when the benzenes are in the trans position (horizontal positions of the lobes), and is open and release the therapeutic molecules (MT) after light irradiation and switching of the benzene nuclei into the cis position (vertical positions of the lobes) (adapted from Lu et al. [LU 08])*

8.2.2. *Nanomachines with mechanical properties*

The idea is to create molecular objects corresponding to certain mechanical components (rotors, gyroscopes, motors, etc.) or to vehicles able to tract, push or to move on a planar surface.

In the past 10 years, a large number of these nanomachines have been designed and an exhaustive description of these can be found in the review article by Kay *et al.* [KAY 07].

Most of these systems are the subject of fundamental and theoretical studies that confirm their feasibility without actually giving rise to concrete applications, the main difficulty being for the moment the creation of network integration in microscopic objects.

In any case, these complex molecular devices require advanced expertise in organic synthesis. Many molecular objects with varied mechanical functions have been created in this way. Nanorotors and gyroscopes, molecular vehicles (nano-wheelbarrows, nano-vehicles), electromechanical nano-switches, artificial muscles, etc., are among the most represented.

8.2.2.1. *Rotors and gyroscopes*

Different rotor models have been imagined. The rotor in Figure 8.12, an "altitudinal" rotor, is made up of a linear and rigid molecule supported by two pillars bound to the surface that keep the rotor suspended and parallel to the surface.

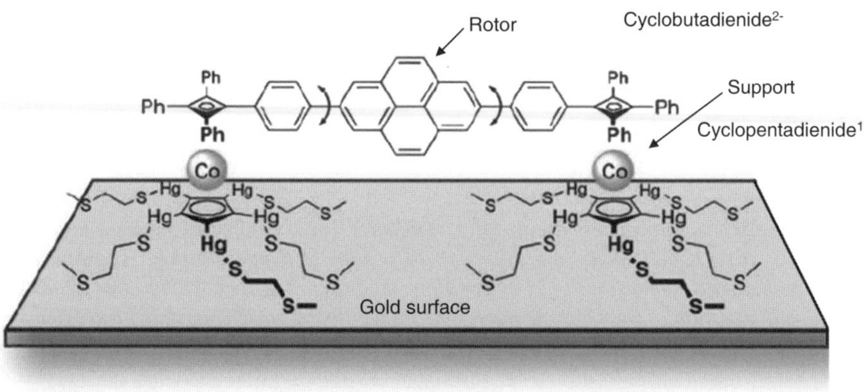

Figure 8.12. *Example of a molecular rotor. The axis of the rotor is supported by two "chandeliers" with five branches bound to a gold surface by sulfur atoms. The assembly of the rotor to the two pillars is performed spontaneously in solution by the addition of cobalt ions (Co) and the formation of an intercalation compound (sandwich compound "cobaltocene") (adapted from Kay et al. [KAY 07], Zheng et al. [ZHE 04])*

The free rotation of the rotor around its own axis is ensured by the presence of single bonds, collinear with each other, bound to the two terminal cycles (cyclobutadienyls substituted by three phenyl (Ph) groups).

The linearity and rigidity of the rotor are due to the alternation of simple bonds and benzene nuclei, the heart of the rotor being constituted of a pyrene nucleus (four condensed benzene nuclei).

The necessity of having two pillars perpendicular to the axis of the rotor is another difficulty that can be resolved, thanks to the formation of a cobaltocene[3], corresponding to the insertion reaction of a cobalt ion between two anionic cycles (respectively cyclobutadienide^{2-} and cyclopentadienide^{1-}).

The creation of such a molecular construction is relatively complex. It has several steps:

– the synthesis of a linear rotor with at the two extremities a cycle with four carbon atoms (cyclobutadienide^{2-}) substituted by three phenyl nuclei that guarantee the stability of the cyclobutadienide anion;

– the synthesis of the pillars constituted of cyclopentadienide anions with its five ended sulfur branches to be bound to the gold surface;

– the assembly of the rotor on the pillars by the formation of the cobaltocene. The cobalt plays the role of assembling element between the two parts of the construction, and also guarantees the verticality of the pillars relative to the horizontal axis of the rotor.

Other models of molecular rotors have been synthesized and studied by several teams worldwide, particularly the creation of asymmetrical rotors, able to be controlled by an external electric field.

Very close to rotors, the construction of molecular gyroscopes has also been envisaged and has given rise to a number of creations. The idea is to synthesize a structure to be able to turn freely inside a molecular "cage" protecting it from the environment.

---

3 Cobaltocenes are sandwich compounds formed between a cobalt cation (Co^{2+} or Co^{3+}) and two anionic carbon cycles. The association of Co^{3+} with cyclopentadienyl^{1-} and cyclobutadienyl^{2-} causes the neutrality and the stability of the whole.

Figure 8.13 describes such an example in which the spinning part of the gyroscope is made up of an iron coordination complex[4] in which it occupies the central position. The axis of the rotor is made up of the alignment of three atoms of phosphorus, iron, phosphorus (P–Fe–P), linked to each other by two single collinear bonds, which give the axis its rigidity and the free rotation. The "cage" that isolates the rotor from its environment is formed by three hydrocarbon chains closed over the two phosphorus atoms and each containing 14 carbon atoms.

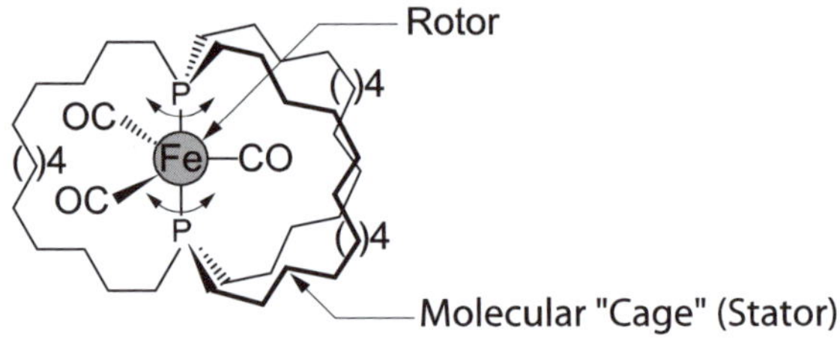

Figure 8.13. *Molecular gyroscope. The rotor made of the iron atom and the three CO ligands is linked to the two phosphorus atoms (P) that play the role of mechanical levels. The rotor can turn freely inside the "cage", formed by three P–$(CH_2)_{14}$–P chains that isolate it from the external environment (adapted from Kay et al. [KAY 07], Shima et al. [SHI 04])*

Other systems have been imagined and synthesized on the same principle, but by introducing an asymmetry in the rotor, which would then allow it act on its own movement, by exerting various interaction forces on the rotor.

The notion of molecular cage is particularly important, as it allows the insertion of the rotor-stator system, not only in a liquid environment but also in different solid environments, like what has been shown in the case of solid molecular crystals (Khuong *et al.* [KHU 07]).

8.2.2.2. *"Motorized" molecular vehicles*

The challenge consists of building molecular constructions that are analogous with macroscopic objects such as wheelbarrows, cars and trucks.

---

4 The central molecular construction is a complex called coordination complex (coordination number 5), with a bipyramidal structure on a triangular base. The iron atom is at the center of the equilateral triangle, and at the summits there are three carbonyl ligands CO. The two last axial ligands (P) are located symmetrically relative to the plane of the triangle and with the iron atom making up the rigid axis of the rotor.

The team led by J.M. Tour at Houston (Rice University) has invested particularly in this research and, as a result of their most recent work, has been able to produce a motorized molecular vehicle, in which light energy is converted into mechanical energy, used to move the vehicle over a surface.

The motor part of the construction is an ethylenic chemical system that uses the *cis–trans* isomerization principle produced by irradiation of the light (see Figure 8.10). Such a system known under the name of Feringa motor (Klok *et al.* [KLO 08]) is characterized by the fact that part of the molecule is flipped over by 180°, which can be used for locomotion (Figure 8.14).

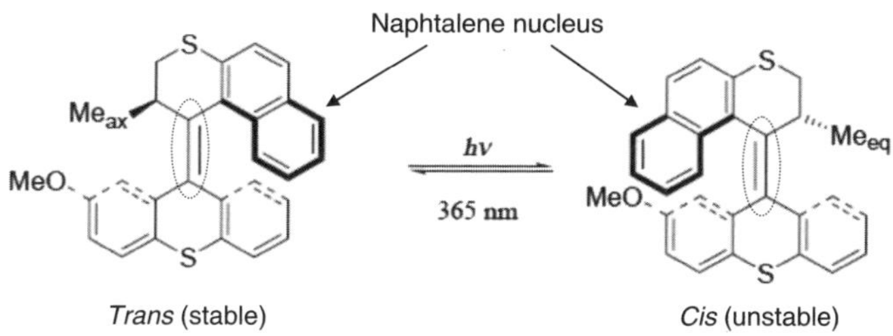

Figure 8.14. *Functional principle of a Feringa molecular motor. Under the effect of light irradiation (λ = 365 nm), the double bond (surrounded by the dotted line) is broken and is formed again with a geometrical change corresponding to a 180° rotation of the naphthalene nucleus that pivots from the right to the left (adapted from Quinn [QUI 03])*

The molecular vehicle imagined and synthesized by Tour integrates this component as the motor element.

The first idea was obviously to use fullerenes as a molecular wheel; this turned out to be inadequate due to their electron acceptor properties that inhibit the *cis–trans* isomerization and then stops the motor from functioning.

The substitution of the fullerenes by p-carboranes (cyclical boron compounds, formula $C_2B_{10}H_{12}$) is, on the other hand, compatible with the *cis–trans* isomerization reaction, and results in adapted "molecular wheels".

A four-wheel "motorized" nano-vehicle with dimensions less than 2 nm has in this way been synthesized by Tour's team (Figure 8.15).

The first exploit taken was to show that a molecular system could be transferred without damage onto a crystalline copper surface in an ultra-high vacuum.

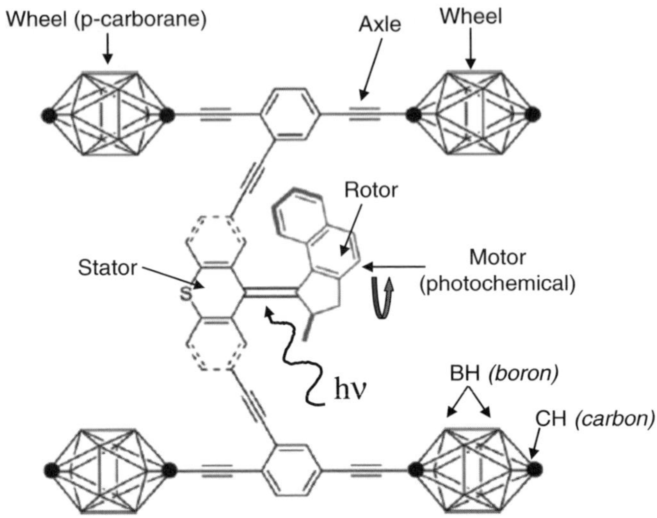

Figure 8.15. *Nano-vehicle with a photochemical motor. Under the effect of the absorption of light (hv), the cis–trans isomerization takes place causing the 180° rotation of the chemical group corresponding to a rotor. Due to friction with the subjacent surface, the rotor causes the movement of the molecular system (adapted from Chiang et al. [CHI 12])*

The observation of movement after light irradiation has not yet been demonstrated, the authors explaining this by the fact that the force exerted by the rotor is less than the forces of the friction of the wheels on the surface. The explanation offered by Tour *et al.* is that the boron atoms that make up the wheels are in a strong interaction with the copper atoms and stop these from turning freely. Surfaces other than copper are being envisaged to decrease the friction.

8.3. Conclusions

The few examples of molecular nanomachines described in this chapter do not claim to give an exhaustive vision of the field, which remains highly prospective. Important efforts must be made to integrate such devices into the microscopic, and ultimately macroscopic, world.

The systems studied are still at a stage of fundamental research, the objective being to develop the molecular mechanisms that govern the movement of these molecular systems under the effect of varied stimuli.

Their elaboration will require advanced expertise in synthetic chemistry and deep knowledge of intermolecular interactions, unique to supramolecular chemistry.

The construction of molecular valves adapted to the stimulated delivery of drugs constitutes a first concrete medical application, which in the future should allow the development of real solutions for the targeted release of drugs in the organism.

8.4. Bibliography

[ASH 98] ASHTON P.R., BALLARDINI R., BALZANI V., BAXTER I., CREDI A., FYFE M.C.T., GANDOLFI M.T., GOMEZ-LOPEZ M., MARTINEZ-DIAZ M.-V., PIERSANTI A., SPENCER N., STODDART J.F., VENTURI M., WHITE A.J.P., WILLIAMS D.J., "Molecular meccano. Part 46. Acid-base controllable molecular shuttles", *Journal of American Chemical Society*, vol. 120, pp. 11932–11942, 1998.

[BAL 00] BALZANI V., CREDI A., RAYMO F.M., STODDART J.F., "Artificial molecular machines", *Angewandte Chemie, Internationale Edition*, vol. 39, pp. 3348–3391, 2000.

[BAL 06] BALZANI V., CREDI A., SILVI S., VENTURI M., "Artificial nanomachines based on interlocked molecular species: recent advances", *Chemical Society Reviews*, vol.35, pp. 1135–1149, 2006.

[CHI 12] CHIANG P.-T., MIELKE J., GODOY J., GUERRERO J.M., ALEMANY L.B., VILLAGOMEZ C.J., SAYWELL A., GRILL L., TOUR J.M., "Toward a light-driven motorized nanocar: synthesis and initial imaging of single molecules", *ACS Nano*, vol. 6, pp. 592–597, 2012.

[DIE 03] DIETRICH-BUCHECKER C., JIMENEZ-MOLERO M.C., SARTOR V., SAUVAGE J.-P., "Rotaxanes and catenanes as prototypes of molecular machines and motors", *Pure and Applied Chemistry*, vol. 75, pp. 1383–1393, 2003.

[ELS 98] ELSTON T., WANG H., OSTER G., "Energy transduction in ATP synthase", *Nature*, vol. 391, pp. 510–513, 1998.

[KAY 07] KAY E.R., LEIGH D.A., ZERBETTO F., "Synthetic molecular motors and mechanical machines", *Angewandte Chemie International Edition*, vol. 46, pp. 72–191, 2007.

[KHU 07] KHUONG T.-A.V., DANG H., JAROWSKI P.D., MAVERICK E.F., GARCIA-GARIBAY M.A., "Rotational dynamics in a crystalline molecular gyroscope by variable-temperature 13C NMR, 2H NMR, X-ray diffraction, and force field calculations", *Journal of the American Chemical Society*, vol. 129, pp. 839–845, 2007.

[KLO 08] KLOK M., BOYLE N., PRYCE M.T., MEETSMA A., BROWNE W.R., FERINGA B.L., "MHz unidirectional rotation of molecular rotary motors", *Journal of the American Chemical Society*, vol. 130, pp. 10484–10485, 2008.

[LIO 09] LIONG M., ANGELOS S., CHOI E., PATEL K., STODDART J.F., ZINK J.I., "Mesostructured multifunctional nanoparticles for imaging and drug delivery", *Journal of Materials Chemistry*, vol. 19, pp. 6251–6257, 2009.

[LIU 05] LIU Y., FLOOD A.H., BONVALLET P.A., VIGNON S.A., NORTHROP B.H., TSENG H.-R., JEPPESEN J.O., HUANG T.J., BROUGH B., BALLER M., MAGONOV S., SOLARES S.D., GODDARD W.A., HO C.-M., STODDART J.F., "Linear artificial molecular muscles", *Journal of the American Chemical Society*, vol. 127, pp. 9745–9759, 2005.

[LU 08] LU J., CHOI E., TAMANOI F., ZINK J.I., "Light-activated nanoimpeller-controlled drug release in cancer cells", *Small*, vol. 4, pp. 421–426, 2008, available at http://www.chem.ucla.edu/dept/Faculty/zink/publications/Small.2008.4.4.421.426.pdf.

[OST 03] OSTER G., WANG H., "Rotary protein motors", *Trends in Cell Biology*, vol. 13, pp. 114–121, 2003.

[QUI 03] QUINN J.R., Synthetic molecular motors, 2003, available at http://www.chemistry.illinois.edu/research/organic/seminar_extracts/2002_2003/Quinn.pdf.

[SAH 07] SAHA S., FLOOD A.H., STODDART J.F., IMPELLIZZERI S., SILVI S., VENTURI M., CREDI A., "A redox-driven multicomponent molecular shuttle", *Journal of American Chemical Society*, vol. 129, no. 40, pp. 12159–12171, 2007.

[SIL 09] SILVI S., VENTURI M., CREDI A., "Artificial molecular shuttles: from concepts to devices", *Journal of Materials Chemistry*, vol. 19, pp. 2279–2294, 2009.

[SHI 04] SHIMA T., HAMPEL F., GLADYSZ J.A., "Molecular gyroscopes: $Fe(CO)_3$ and $Fe(CO)_2$–$(NO)^+$ rotators encased in three-spoke stators; facile assembly via alkene metatheses", *Angewandte Chemie International Edition*, vol. 43, pp. 5537–5540, 2004.

[STE 98] STEINBERG-YFRACH G., RIGAUD J.-L., DURANTINI E.N., MOORE A.L., GUST D., MOORE T.A., "Light-driven production of ATP catalysed by F0F1 ATP synthase in an artificial photosynthetic membrane", *Nature*, vol. 392, pp. 479–482, 1998, available at http://www.qi.fcen.uba.ar/materias/tfqsb/papers/atp-con-luz.pdf.

[STO 99] STOCK D., LESLIE A.G.W., WALKER J.E., "Molecular architecture of the rotary motor in ATP synthase", *Science*, vol. 286, pp. 1700–1705, 1999.

[VAL 00] VALE R.D., MILLIGAN R.A., "The way things move: looking under the hood of molecular motor proteins", *Science*, vol. 288, pp. 88–95, 2000.

[YAN 12] YANG P., GAI S., LIN J., "Functionalized mesoporous silica materials for controlled drug delivery", *Chemical Society Reviews*, vol. 41, pp. 3679–3698, 2012.

[ZHE 04] ZHENG X.L., MULCAHY M.E., HORINEK D., GALEOTTI F., MAGNERA T.F., MICHL J., "Dipolar and nonpolar altitudinal molecular rotors mounted on an Au(111) surface", *Journal of American Chemical Society*, vol. 126, pp. 4540–4542, 2004.

Conclusions and Outlook

Since the famous conference held by R. Feynman (1959), which received skepticism at that time, and during which he advocated the study of infinitely small structures, the importance of nanotechnology has been well acknowledged. Scientists are convinced of the technological advances to come in fields as varied as electronics, energy, mechanics and medicine. According to them, these technological advances will come as a result of advances made in nanoscience and nanotechnology.

The successive reports on dedicated investment by public bodies across the world confirm this progression. There are some predictions that in 2015 a turnover relative to nanotechnology could reach between $500 and $1,000 billion at least, with $2,500 billion being a more optimistic figure (A. Hullmann, 2006, http://cordis.europa.cu./nanotechnology).

This outstanding progression is due in part to the incessant improvement of the observation methods of the infinitely small, in particular the invention of near-field microscopes in the 1980s that revolutionized the study of nanomaterials, and also to the discovery of carbon nanomaterials (fullerenes, carbon nanotubes and most recently graphene), that every day provide new application and properties, sometimes unexpected.

Parallel to these discoveries, chemical engineering has also reached its utmost level of development, enabling the customized design of nanoparticles from a wide range of materials and then organizing them into regular networks, which are the source of new electronic and optoelectronic properties. It is also a new way of structuring surfaces on a subnanometric scale, with the objective of replacing current microelectronics by molecular electronics. This is still far from large-scale production, and many conceptual and technological obstacles need to be overcome before concrete applications are made.

However, in the domain of energy, and, in particular, in the process of electrochemical storage and the conversion of solar or chemical energy into electricity, the use of nanomaterials in a variety of forms is becoming significant; undoubtedly, this has led to great improvements in energy yields.

The large-scale production of these nanomaterials, no longer mere lab toys, is now a common occurrence. The cost of production has decreased considerably, while keeping the degree of purity high; this has led to the prediction of massive uses over the next decade in a multitude of production processes of various components, especially in chemical conversions that require finely divided catalysts.

Their use in medicine has already given impressive results, as much in diagnosis as in therapy. Efforts are being made to put new strategies in place and for the handling of nanoparticles. Their functionalization by molecules with therapeutic properties and the understanding of their mode of action after injection into the body, in particular the increasingly precise targeting of drugs, are indeed significant. Considering the rapid progress made in this domain, we will see in the near future spectacular breakthroughs in the treatment of diseases such as cancer.

The study and the creation of artificial nanomachines, which can move around or perform translational and rotational movements, is part of a very prospective domain of research, whose future aim is to imitate certain biological molecular machines. Recent designs are complex though impressive. However, their integration into microscopic objects or biological systems is still underdeveloped, apart from the molecular valves that were recently tested in the targeted and controlled delivery of drugs.

The increasingly generalized production and the use of nanomaterials evokes the issue of their toxicity. A controversial topic, currently it is difficult to get a precise idea of the extent of the danger caused by nanomaterials due to the diversity of the different protocols used to evaluate the risks. In any case, a policy of precaution is required, but it is first necessary to define the physicochemical parameters that can be treated as the cause of pathological disturbances.

Index of Names

L, M

O, P

R

S

T, U, V

W

Index

M

N

O

U, V

W, X, Z